PRAISE FOR *THE FUTURE OF SCIENCE IS FEMALE*

"As Geena Davis says, 'If she can see it, she can be it.' By focusing on women working to change the current world, Zara Stone has created an important book. *The Future of Science Is Female* will serve as inspiration—not only to young girls needing role models in science and technology, but to anyone who refuses to let society dictate what they 'should' be doing. I thoroughly enjoyed reading and learning about the stories of these diverse ladies who will help to create our future."

—ALICIA MALONE, AUTHOR OF *BACKWARDS AND IN HEELS* AND *THE FEMALE GAZE*

"If we want to encourage more girls to enter into science careers, we need to show females that this field was made for them to conquer. Zara Stone's engaging narrative does just that by introducing readers to brilliant but relatable female role models who are using science to positively change our world."

—COLLEEN RUSSO JOHNSON, PHD AND CHIEF SCIENTIST FOR DREAM STUDIOS INC.

"*The Future of Science Is Female* shows us that the future is happening now! Zara Stone will make you want to grab your lab coat and join the women making scientific her-story!"

—KELLIE GERARDI, AUTHOR OF *NOT NECESSARILY ROCKET SCIENCE*

The Future of Science Is Female

The Future of Science Is Female

The Brilliant Minds Shaping the 21st Century

Zara Stone

Coral Gables

Published by Mango Publishing, a division of Mango Media Inc.

Cover Art and Interior Layout Design: Jermaine Lau

For permission requests, please contact the publisher at:
Mango Publishing Group
2850 Douglas Road, 2nd Floor
Coral Gables, FL 33134 USA
info@mango.bz

For special orders, quantity sales, course adoptions and corporate sales, please email the publisher at sales@mango.bz. For trade and wholesale sales, please contact Ingram Publisher Services at customer.service@ingramcontent.com or +1.800.509.4887.

The Future of Science Is Female: The Brilliant Minds Shaping the 21st Century

ISBN: (p) 978-1-64250-319-7 (e) 978-1-64250-320-3
BISAC: YAN006140, YOUNG ADULT NONFICTION / Biography & Autobiography / Women
LCCN: Has been requested with the Library of Congress Cataloging

Printed in the United States of America

Please note some names have been changed to protect the privacy of individuals.

TABLE OF CONTENTS

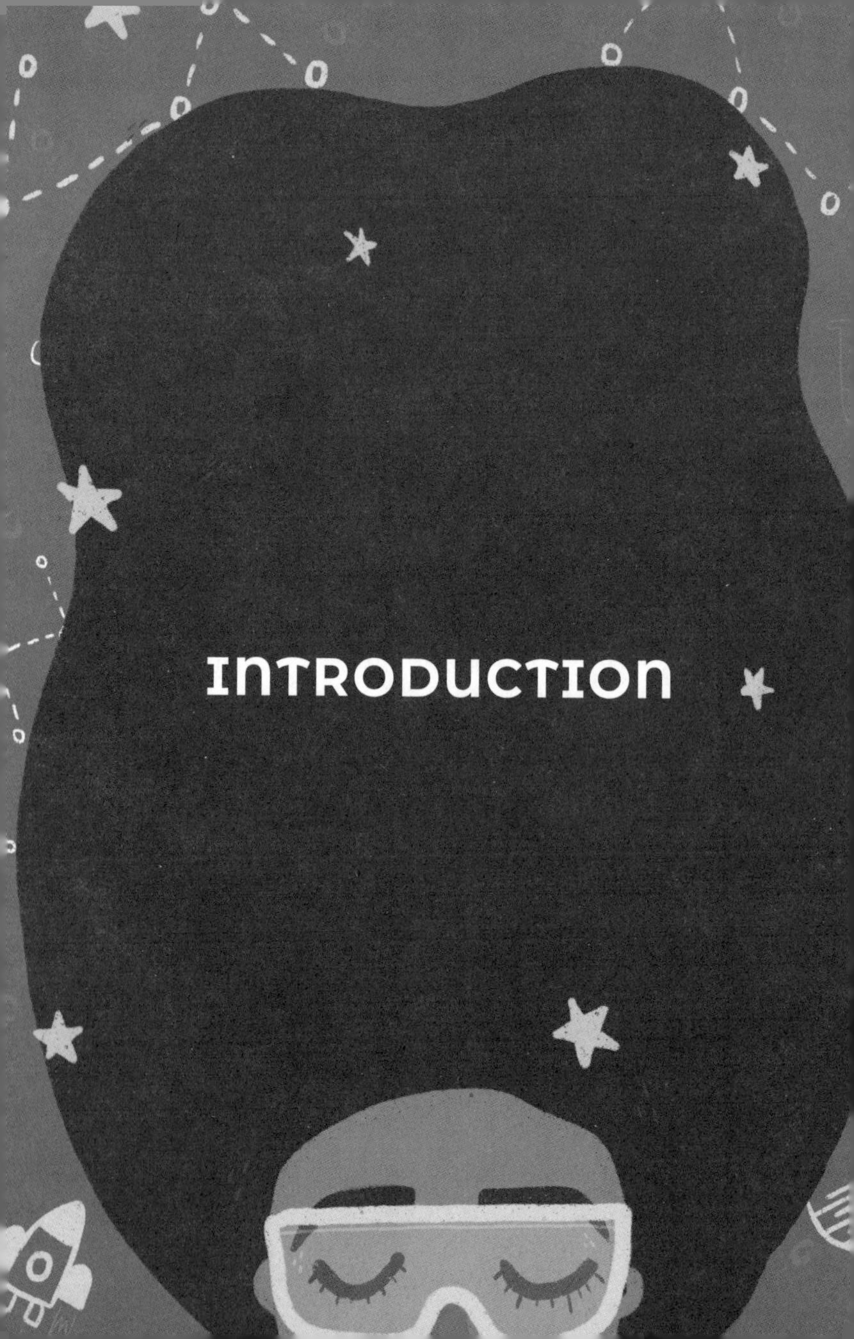

INTRODUCTION

Throughout history, badass women have created world-changing advances in science and technology, and right now they're finally getting the recognition they deserve. #preach

In San Francisco, scientist Etosha Cave developed a machine that sucks carbon dioxide out of the air and turns it into useful plastics and fuel when she was twenty-nine. In Burlington, Canada, seventeen-year-old Riya Karumanchi built the SmartCane, which uses computer vision, haptics, and GPS technology to steer visually impaired people around obstacles. The SmartCane's vibrations also provide directions. In Connecticut, twenty-nine-year-old Ashley Kalinauskas's company, Torigen, creates and sells a working cancer vaccine for dogs. Thanks to Ashley's dedication, hundreds of puppies' lives have been saved. The list of amazing women in the world *right now* goes on and on. These scientists and technologists are Black, queer, Asian, disabled, Latina, and white.

We need more of them.

Today, fewer than 4 percent of Latinas and 3 percent of Black women get doctoral degrees in science and engineering. Women make up 20 percent of all undergraduate degrees in engineering, physics, and computer science, but only 11 percent end up working in STEM.

Implicit bias, sexism, and a culture that recommends Bratz dolls to girls and Lego kits to boys play a part in this. Then there are the history books full of white men, which suggest that women aren't as important or as clever as dudes are. The news is full of stories about cool tech and science startups and tends to focus on the Evan Spiegels, Elon Musks, and Mark Zuckerbergs of the world. When they mention female CEOs, they tend to have fashion- or beauty-related businesses.

We need to know about the Etoshas and Riyas and Ashleys of the world.

The more you hear about people who look and sound like you and are doing great things, the easier it is to realize that you can do that too. I wrote this book so girls and gender-nonconforming and nonbinary people everywhere can learn about people who dealt with similar problems and how they refused to let that rule them.

Why This Matters

In 2007, I interned at a British men's magazine that covered celebrities, fast cars, and the latest gadgets. It was a competitive internship, and I was excited to

work there. I really wanted to make it in journalism; the idea of telling people's stories and learning about the world seemed like the best job ever. I was one of two girls in the office.

I did everything that people told me would make me a success: arriving an hour early and leaving late and saying yes to everything they asked me to do, from photocopying to typing up notes. I loved the energy of the place. Early on, I signed for a delivery of products from a big electronics brand. There were so many that the editor, a fifty-something white man, said everyone should take one to review—including me. The pile included headphones, a camera, fitness trackers, and more. He asked me what I wanted, and I said I'd love to review the camera. He frowned. "Why not take the electric toothbrush or the hairdryer?" he said. "I think you'd be good with them."

I nodded—I was trying to be the *best intern ever*, remember?—and took the electric toothbrush home. A week later I turned in a kickass review, which went in the magazine. Then they gave me a hair straightener to review. A pair of smart scales. A pink laptop. I loved learning about technology, but I hated being put in a box.

I've never forgotten that editor. Maybe he genuinely thought those gadgets were best for me. Maybe he didn't realize how sexist and belittling it felt to be the girl with the pink technology. But so what? It still stung, and it influenced what work I did there and how other people looked at me. His microaggressions stuck with me.

I went on to have a great career in journalism, with a focus on the intersection of culture, technology, and science. I worked as an on-air reporter for ABC News and as a writer for the *Washington Post, The Atlantic*, *Wired* magazine, the BBC, and more. I've interviewed Steve Wozniak, DJ Tiësto, Cat Deeley, Spice Girl Emma Bunton, Watchmen designer Dave Gibbons, firefighters, mermaids, and more. But I got here in spite of, not because of.

These kinds of microaggressions still happen. In schools across the country, about half as many girls as boys are interested in STEM by eighth grade. That drops to 15 percent by the end of high school. That's not cool. The world is getting techier and techier, and we need women to be part of its creation. If you're not in the game, you don't get a say in what it looks like.

The Takeaway from This Book

Each chapter in this book examines a different world problem, from climate change to the future of work, and introduces the female scientists who are working on solutions. You'll learn how they got to where they are and what challenges they faced and overcame along the way.

It provides an overview of some of the coolest and most exciting science and technology projects happening today—all pioneered by badass women who challenged the status quo. They saw problems that needed solving and wouldn't take no for an answer.

You'll learn the fascinating, complicated stories of how this diverse group of women got started—from the perspective of those still working it out as they go along. Forget the ivory tower of accomplishment; learn about the everyday drama, tears, and adventures these awesome ladies face as they race to fix everything that men f***ed up.

This book isn't going to turn you into a scientist. It won't make you the next Steve Jobs or Bill Gates or Sheryl Sandberg or Greta Thunberg. But it will show you that way more women than you thought are working in these fields and that they're involved in awesome and incredible projects—some which seem like they're lifted straight from Hogwarts. #slay #blackgirlmagic

In the words of Beyoncé, "Who runs the world? Girls! Girls!"

You can be that difference in the world.

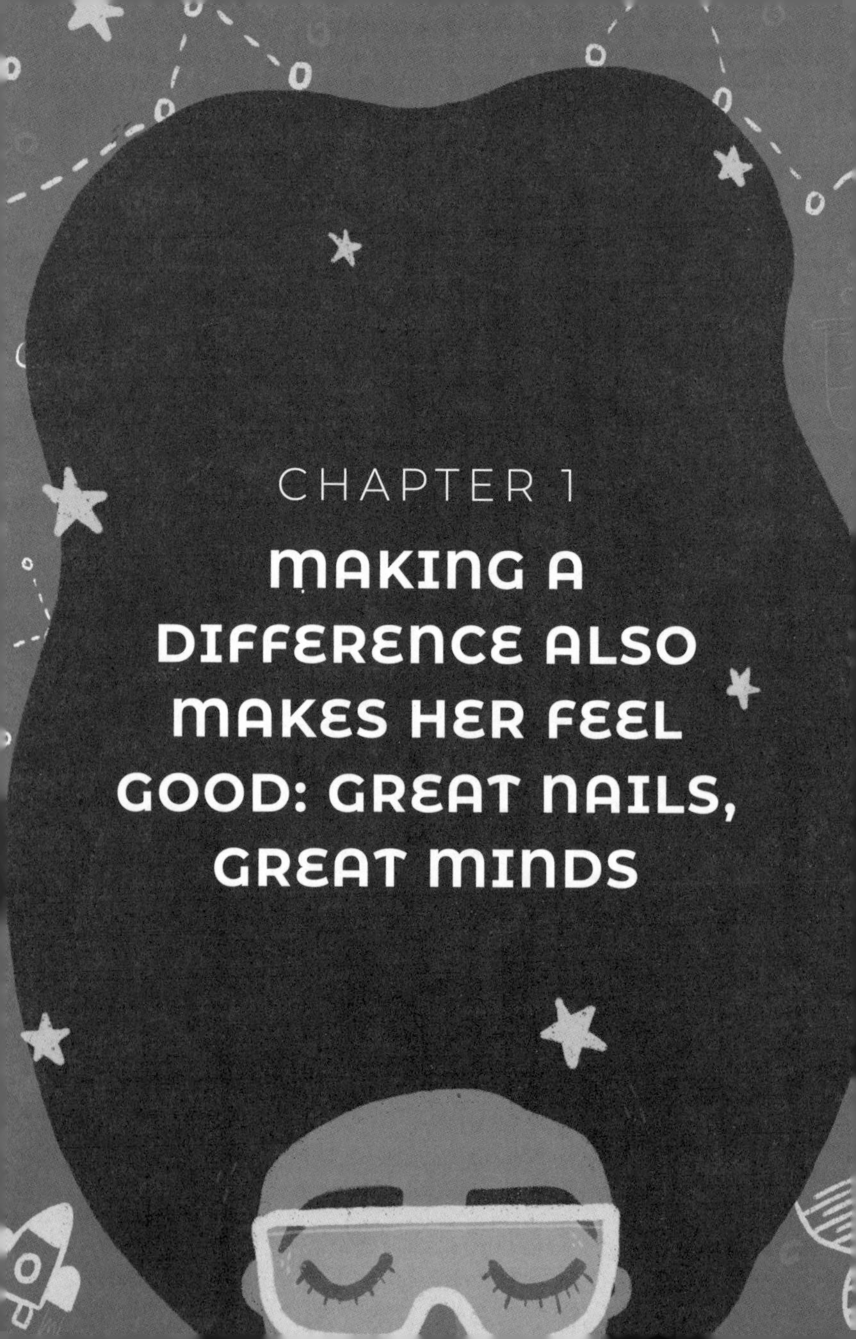

CHAPTER 1

MAKING A DIFFERENCE ALSO MAKES HER FEEL GOOD: GREAT NAILS, GREAT MINDS

Pree Walia has really great nails, and she knows it. Today, one of her digits features a pizza emoji, another has the poop emoji, and a third has a crystal-clear image of Maverick, a fluffy brown sheepdog she considers the office mascot. "At my core, I'm a girly girl," she said, pushing her waist-length brown curls behind one ear.

We meet in November 2019, inside Star Space, the Silicon Valley coworking center she works from. Walking through the space is like entering a CB2 catalogue: pastel colors, velvet sofas, and a hanging chair swing. "You might recognize this from our videos," she said—the place is pretty enough to use as a backdrop for her company promotions.

On social media, Pree's all about the fun times—her feed features beautiful sunsets and international travels, and she's pictured in cute dresses, hanging with friends and swigging White Girl Rosé. With her love of fashion and beauty, Pree stands out among the tech bros in their Patagonia fleece-lined vests, joggers disguised as jeans, and AllBirds sneakers. The fact that she's a girl also sets her apart; in

Silicon Valley, men hold around 80 percent of all technology jobs.

It's about the worst possible place to found a beauty startup, even when it's a technology *and* beauty company. Working in beauty wasn't in her life plan, but the beauty space—nails, in particular—has been her world for the last five years.

Pree didn't pay a manicurist for the cheeky pizza emojis and puppies on her digits, and she didn't paint them herself. Her tight-looking talons are courtesy of the Nailbot, a portable printer that prints nail art on her fingers. Designs are chosen via her cellphone app. She knows what she likes. Growing up, she had every manicure in the book—French tips, gel manicures, glitter, stripes, geometrics, with talons every color of the rainbow. "I like being a girl—whatever that means, [even if] it's a socially constructed version of femininity," she said. "I like getting my hair done. I like getting my nails done. I like getting spa treatments. They make me feel good."

Making a difference also makes her feel good, which was why she founded her company, Preemadonna, with the Nailbot as her first product.

Sure, great nails won't get you to the White House (even if they might help, just a little)—but if that's

all you see when you look at the Nailbot, you're not thinking big enough.

"Our big vision: can you learn how to code nail art?" says Pree. She views the Nailbot as a stepping stone into the larger world of STEAM (science, technology, engineering, art, and math). "The nature of this product is that it's a vehicle for more: artists, hackers, coders, and programmers!" Yeet.

So yes, your nails will look sweet, but that's just the beginning. Learning to build a touch-screen printer and coding in your nail designs will get you far in life. The idea is to use the Nailbot to get male, female, nonbinary, and trans people interested in technology. Once she has their attention, she introduces them to STEM initiatives and the MakerGirl program, which teaches 3D printing and STEM to girls ages seven through ten.

Here's Why This Matters

For better or worse, the world runs on code right now; it's how traffic lights change from red to green, how food gets from the field to our plates, and how medical robots perform a bunch of invasive operations. It's how stores decide what dresses to restock and how schools evaluate your grades.

But most of this code is written by dudes. In 2018, women made up around 21 percent of the global tech workforce at Google and Facebook, the companies essentially running the world today. In 2014, this stat was worse; women made up 15 percent of tech jobs at Facebook. For women of color, it sucks more; in 2018, 0.8 percent of female tech hires at Google were Black and 1.4 percent were Hispanic. At Facebook, the numbers are so low they didn't separate them by gender; 1.3 percent of all tech hires were Black, and 3.1 percent were Hispanic.

Working in technology is about more than writing the code that runs the world. Coders also get big salaries, big perks, and big power. They matter in the world order. So the fewer women there are on the payroll, the less it looks like women matter.

Sure, in the olden days, women stayed at home and men went to work. But people also had smallpox and polio!

Today, most schools run coding classes and special programs, so all genders get a shot at learning these skills. In many schools, they're required subjects; if you fail Java 101, you have to retake it. But—and it's a big but—while most everyone now learns to code in school, that doesn't mean they code at home.

One study found that 40 percent of boys who learn to code at school will code at home for fun, compared to 5 percent of girls.

One reason for this is what's available to code. Not every girl wants to mess around with Star Wars or Batman. Many do, but, for some, they lack any appeal. There are some gateway coding toys such as GoldieBlox (more on this later), but none approach coding from a beauty perspective. That's where the Nailbot comes in.

* * *

As a little girl, Pree always thought her career would be in politics. She was born in the South, in New Orleans. Her family moved to Madison, Mississippi, when she was nine years old. No matter where she lived, she'd plunk herself down in front of the TV whenever the debates were on, sitting as close as she could to the screen to take in everything.

One of her earliest memories is of watching the debates with her dad. He'd get so excited watching the various politicians speak. Her dad loved America, loved it with the type of passion that only an immigrant has. He'd emigrated to the US from India, during one of the many wars in the 1960s, to attend university and find a better, safer life.

America was everything he'd hoped for; he was hired as an electrical engineer by a great company, and Pree's mom ran a bunch of successful franchises—everything from a Baskin-Robbins (Pree's top tip: try the Gold Medal Ribbon flavor) to jewelry store chains. In this country, you can do anything, be anything, he told Pree. It's the land of opportunity!

Maybe I'll be a senator one day, she thought. *Or even the president!* At school she took speech and debate, working on her diction till all traces of her Southern accent were gone. She signed up for student council and organized the school's philanthropic drive. Her favorite thing was getting people together, all working toward a bigger purpose. She craved the feeling of a task successfully completed. That feeling of achievement was addictive. In high school, she interned for the Mississippi Secretary of State, helping out on charity enforcement.

"I was fascinated by the political process," she said. But Pree realized that being in charge wasn't what drove her. "Whether you're heading student council, running for president, or being an entrepreneur, it's about having a vision of the future."

Pree and her two elder sisters (and sometimes her little bro) helped out with her mom's businesses; many nights, she'd scoop ice cream for the local kids

and listen to their chatter. So many different people, so many different wants. The world was so big, and there were so many things she could do.

After high school, she studied history with a gender studies minor at Northwestern University. College life was awesome; she joined a sorority, made lifelong friends, and enjoyed feeling like an adult.

But Pree found that she couldn't stay away from politics. In 2003, she volunteered for John Kerry's presidential campaign (spoiler: he lost), and when she graduated from college in 2004, she was hired by Campaign Corps, a program at Emily's List that supports women that are running for office. For an entry-level graduate, the program worked a bit like Teach for America; after learning how to operate a campaign, Pree was sent to a congressional race in Arizona. She followed this with a stint in Washington DC, working for the Democratic Congressional Campaign Committee, then headed by Nancy Pelosi. Next, she headed west to work on the California primary for governor. She was twenty-four years old.

Still unsure about what to do with her life, she applied to business school, hoping that would narrow it down. Maybe consulting? That would give her variety and a chance to keep affecting large groups of people. She applied to all the big-name firms...but no one would hire her. "I think that with

my crazy energy and my crazy background, I wasn't the best fit for them," she said. She confused the companies; a history major with a background in political campaigns and ice cream who wanted to get into consulting? Frustrated, Pree worked on a new plan. She knew that she valued a high-energy workplace that operated at a fast pace. Maybe joining a startup?

In 2009, she was hired part-time by a San Francisco Bay Area startup that specialized in advancing LED light systems; they focused on connected technology and color-changing bulbs. She was the only woman on the team. But the world was changing around her. By the time she left, almost a fifth of the employees were women.

* * *

The tech landscape was changing in other ways as well. In 2011, Debbie Stirling, a twenty-eight-year-old designer at a San Francisco branding firm, was inspired by a chat she'd had with her friends during brunch. They were reminiscing about the favorite toys they'd played with as kids when one woman complained that she'd had to borrow her brother's Legos, as they were never bought for her. Debbie knew that feeling well. She'd grown up in Rhode Island with the idea that "engineering" was a nerdy

and intimidating word. Her parents wanted her to be an actress. They bought her Barbie dolls, but no Legos. Engineering was just for boys, she thought. "I was so wrong!" Despite having no construction toys as a kid, with a teacher's guidance, she'd turned to science, and graduated from Stanford University with a degree in engineering and product design. "I became obsessed with the idea of getting girls interested in STEM through toys," she told reporters.

For the next nine months, after she finished work, she pored over research about how kids' brains develop, and how the different genders learn. "My aha moment: realizing that combining storytelling with building was a way to get girls interested in STEM," she said. She'd discovered that young boys have good spatial skills (one reason that they love Legos so much) whereas young girls have great verbal skills; they're all about the characters and the stories.

Debbie named her company GoldieBlox, targeting five- to nine-year-olds with a book-and-build combo. The kit contained a construction set and a book, starring Goldie, a blond tween inventor who loved purple overalls and taking things apart.

Whenever Goldie had a problem, she built a machine to fix it. The set let kids build that too, unknowingly learning basic engineering principles

as they followed her adventures. Debbie's color palette was light, fresh, and fun—there is pink in there, but it hasn't been "pinkwashed." After Goldie, she introduced Ruby Rails, an African American coding whiz.

Excited with her design, she brought her prototype to the New York Toy Fair, the best-known toy fair in America. No one liked it. "I was told that toy patterns were innate—girls liked playing with dolls and boys liked building," she said. They told her it was a noble cause, but it would never go mainstream. She disagreed. "I knew these were outdated stereotypes that needed to change."

She wasn't going to quit now; she believed so much in the project that she'd spent her life savings on it. So she turned to Kickstarter. "GoldieBlox is to inspire girls the way Legos and Erector sets have inspired boys to develop an early interest and skill set in engineering," she wrote. "It's time to motivate our girls to help build our future."

She offered rewards to people who contributed, including a yellow T-shirt that read "more than just a princess," with the *O* replaced by a machine cog. Her video went viral, proving what she already knew; kids were crying out for this. In 2014, GoldieBlox's ad aired during the Super Bowl, and, in 2019, the company

was valued at more than forty million dollars. Debbie wasn't the only female entrepreneur making waves.

In 2014, engineer Sara Chipps and fashion guru Brooke Moreland decided to team up. The longtime friends launched Jewelbots, a plastic friendship bracelet that teaches girls to code. A flower adorned each Jewelbot bracelet, which was fully programmable; you could set it to react to your friends, buzzing when Bryony was near and changing color if it picked up Margie over Bluetooth, for example. It could also send secret messages to pals, and the open-source app allowed users to create their own programs, such as class schedules or texting their parents if they felt unsafe. "Girls are not one-dimensional," Chipps told reporters. "We want to show them that you can be interested in tech and everything else that's fun about being a girl."

Then there's Roominate, founded by twenty-three-year-old Alice Brooks and twenty-five-year-old Bettina Chen, two engineering undergrads, who'd come up with a novel take on the traditional dollhouse—build it yourself. Kids didn't just have to construct it, they had to wire it; the kit included circuit boards to power lights, fans, and working elevators. They displayed it on *Shark Tank*, where

Mark Cuban gave them $500,000 for 5 percent of their business.

This is a big deal. There's a lot of research that shows girls lose interest in science by age eight. By eighth grade, about half as many girls as boys are looking into STEM careers. The numbers suck. 75 percent of girls start out interested in STEM, but that drops to 15 percent by the end of high school. This keeps going; only 12 percent of women complete STEM degrees, and only 25 percent of those work in STEM ten years later. Changing this involves changing the way girls think about STEM, and the culture around it.

* * *

In 2011, Pree was vaguely aware that these companies had started. They seemed really cool, she thought; why weren't they around when she was a kid? In 2012, the same year that GoldieBlox launched, Pree graduated from business school. With no commitments, she bought a one-way ticket to London. Then she flew to France. Then Spain. "It was a coming of age [adventure]," she said. She didn't know what she was searching for. Maybe a fairytale romance with a European? A job at Google? "I wanted to know how I could make a contribution to the world," she said.

In Spain, one of her classmates was getting married. She had her dress and her shoes, but her nails looked raggedy, all chipped and dull from flying. *Time for a manicure*, she thought. But she couldn't find a salon—every place wanted you to book well in advance, and they were really expensive. The nail bars so common in New York and LA and Houston were nowhere to be seen. "That was my aha moment," she said. "It shouldn't be so hard to get a simple manicure."

She flew back to the states at the end of summer and took a full-time role at the LED startup in California. But her nail idea, still unformed, bugged her. She kept coming back to it, over and over again. *What we need is some kind of portable nail salon*, she thought. She'd never worked in fashion or beauty. All she knew was politics and connected hardware. *Maybe that's a good thing, a weird advantage*, she thought; she knew how to work with people in Silicon Valley, how to raise money, and how to get people interested in what you were saying.

She kept refining her idea. The travel nail salon idea was quickly trashed, but she kept returning to the idea of a printer. She wanted her startup to be more than just nails. Would people want pretty nails enough to learn how to build a printer? But why? She drilled down to the idea of a nail art printer.

But she needed help. Looking through LinkedIn, she kept returning to Casey Schulz's profile; Casey was a systems engineer with a background in inspiring girls with cool, creative technology. Pree messaged her, and they met up at the coworking space where Casey taught kids how to use Arduino. They clicked. "I want the Nailbot to be the first in a long line of products," Pree told her. "The point is pretty nails plus learning design software, coding, and more." Casey was in.

Now Pree needed to decide on a company name. She thought hard about that. What did she want her company to be called? How would it represent her and her mission?

She didn't want to go childish or stereotypical, no using words like *princess nails* or *fairy prints*. The word *printing* gave her pause. Saying it slowly, out loud, it sounded like her name—Pree-nting. *Prima donna*. Preema donna! She liked it. "We're taking it back!" she said. Sure, the term *prima donna* has connotations of being difficult, but that was entrenched sexism. She's reclaiming it for herself. "The original prima donnas were these very talented women, the stars of the show," she said. "People want to consider them difficult, but... Look, if you're a CEO, that's 'visionary' of course, you're going to be difficult. I think the best leaders are difficult because

they're always pushing you. You're going to stand for something."

Next, the duo turned their attention to money. Casey was working her day job, so it was up to Pree.

But to reach people, they needed some cash. *That shouldn't be too hard*, she thought. The Valley invested huge amounts in silly startups every day, such as $118 million for a juicer that wouldn't juice, and $20 million on a startup that made your emails smell. For real. Plus, the nail market was booming; in 2013, 92 percent of tweens and teens used nail polish. 14 percent used it daily. This would be a breeze.

She approached venture capitalists for funding. She applied at a lot of technology accelerators. But nobody was biting. Most VC firms and tech accelerators were run by (mostly) white men, who didn't get the point of her company. She'd go to meetings and demonstrate the Nailbot, and they all refused to get their nails done by it. It was demoralizing. Every day, she woke up, moved from her bed to the couch, and waved her roommates—two friends from business school—goodbye as the headed out to their well-paid banking jobs. They were supportive of her, but Pree was losing hope. "My dad was like, is the finger painting over?" she said. Her bills were mounting up. Living in San

Francisco was expensive, especially when you had no cash coming in.

She messaged her freshman roommate, Diane Donald, to talk about how tough everything was. "Don't give up," Diane said. "I know you can do this." As a mother of three, Diane really liked Pree's idea. She was Pree's first investor. "I'm probably going to lose all my money, but I believe in you," Diane told her.

Hmm, thought Pree. Silicon Valley didn't want to fund her, but maybe she was looking in the wrong places. She messaged her old sorority, Tri-Delta. "This is what I'm working on," she told them. "What do you think?" They loved the idea and sent a check. Little by little, funding trickled in, enough to get her to a working prototype.

She felt exceedingly grateful. "This company exists because of the people who believed in me—my sorority sisters, my friends, my mentors," she said. Helen Greiner, the creator of the Roomba vacuum, also invested. The difference, Pree realized, was that she'd been trying to get money from old white men. But the women she approached saw the value. Today Preemadonna has an all-female millennial board.

To see how tweens and teens and early twentysomethings felt about the Nailbot, Pree hosted stealth parties. The first few were filled with her friends' kids and their friends. They went really well—everyone was excited about the Nailbot and curious as to the tech behind it. "Can I intern for you?" asked one. "Can I be an ambassador?" said another. Sure, she said.

She developed those parties into a mini-TED talk about building and developing a product, which she presented in high schools. The kids were sharp, she said; in the Q and A, they asked her what language she'd coded the app in and how hard it was to get investors. She ended her session with free manicures for everyone! She gave the same talk in many venues, including Kode with Klossy camps, Maker Girl events, and the Girl Scout convention. "It's not where a traditional beauty company would go," she laughed, "but it resonated with me as a founder." Her ambassador network grew.

In 2015, her hard work paid off and she was accepted into the Hax hardware accelerator, which provides companies with $100,000 and easy access to experts and mentors. They're famous for launching the world's first connected tampon and the Dispatch delivery robot (which was bought by Amazon). At Hax, Pree and Casey learned how to use the iPhone's

front-facing camera on the inkjet. Then they teased their product at TechCrunch Disrupt, a San Francisco startup showcase that launched Everlywell's home lab testing kit and fertility startup, Future Family.

Excited by all the positive feedback, Pree ran an Indiegogo campaign in 2016. She did everything wrong. She priced the Nailbot at $199, but the product wasn't ready to ship, and she had no ETA for buyers. She didn't set up a waitlist. "By all measures, that Indiegogo campaign was a failure," she said—with one upside. Her campaign went viral and was featured on most news sites in America. "We got our word out, and we paid no money for marketing!"

She kept iterating, based on feedback from her ambassadors. Now, when you print on your nails, your phone automatically captures a picture of them and lets you share them straight to social media. She designed a bunch of skins for the printer itself; girls told her that they wanted to personalize the machine's color.

The process is super quick. First you paint your nails with their primer, followed by any white nail varnish (she supplies these with the kit), then select a picture from the app, and use the front camera on your phone to visualize the design over your digit. Press a button, and whoosh—five seconds later your

nail is printed, and dry. "The only limitation is the canvas of your nails," Pree said.

There are thousands of emoji-style images in the Nailbot library, with more added every day. She's collaborated with nail influencers and artists to create unique designs for the app, and users can choose to learn Adobe Illustrator or Photoshop (she has online how-to videos), then upload their own designs as well. "Just like you can share a Spotify playlist, you can share a nail art playlist!" Future iterations, teased on her Instagram, show more

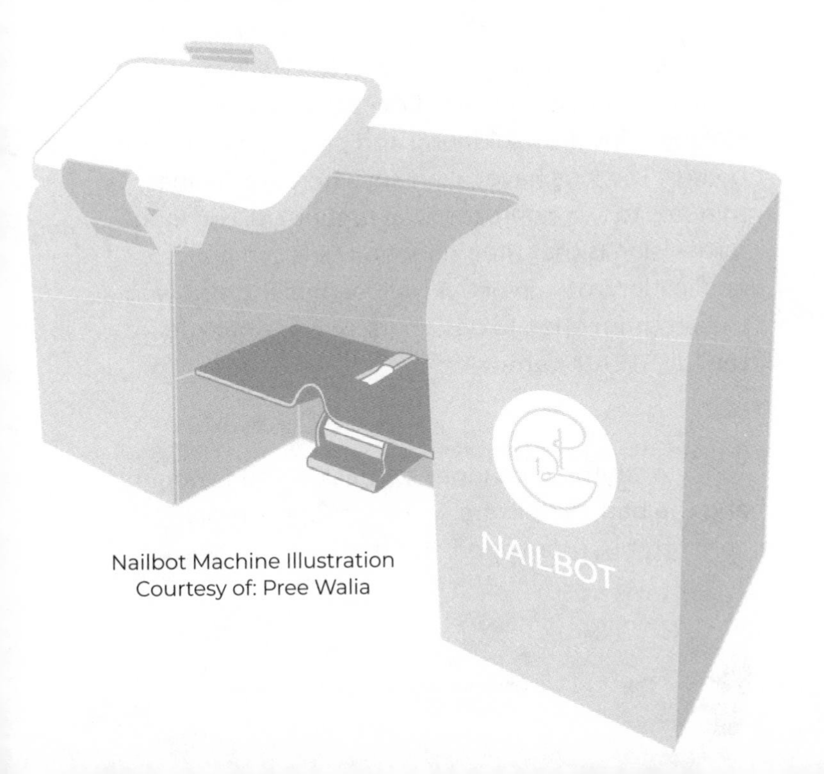

Nailbot Machine Illustration
Courtesy of: Pree Walia

complex designs printed on blue and purple nails; version two, perhaps? She smiled but wouldn't confirm. But augmented reality (AR) is the works, she said. "You can print any emoji or any photo from your camera onto your fingernail," she said. "But it's not a toy. It is a beauty tool, but for girls, it's also a learning tool."

When "Preemadonnas" really dig the tech, she links them with Maker Girl Mentors, where she's on the board. They encourage girls to get into tech and leadership roles in STE(A)M with mentoring and career help.

Pree beta-tested her build-it-yourself Nailbot maker kit with a number of her ambassadors. Making it involves soldering and playing with circuit boards. The kids have taken anywhere from ninety minutes to five hours to make it, she said. Her long-term vision is that Preemadonna girls can build applications on top of the Nailbot, but she also wants the program open to designers, makers, hackers—the full STEAM gamut.

* * *

Today, everything changes so quickly. The good and the bad. People are more accepting, more comfortable, more open. But then there's the

growth of online trolls, the dark side of the internet, and people using tech for bad.

We need to look back to learn how far we've come. Take the 1992 Talking Barbie doll, dressed in a rainbow rockabilly skirt, with crimped hair and a tie-dyed denim jacket. The user pressed a button on her back and she chirped out her lines. "Math class is tough! Party dresses are fun. Do you have a crush on anyone? Math class is tough!" Even in those dark ages, this was not okay, and she was quickly recalled. But it took Barbie a long time to get with it—over the years, her careers have included working as a model, a ballerina, a McDonald's server (fo' real), and Miss America.

She joined the twenty-first century in 2012 with the release of computer engineer Barbie, complete with pink glasses and a pink laptop, and in 2017 the Barbie drone was a Christmas sellout with kids. Today, there are Barbie-branded online coding classes, AI interfaces in Barbie's Dreamhouse, and a Barbie coding curriculum for schools. Their message: it's okay to code and be cute, if that's what you want.

Caring about how things look used to be considered a vain, "girly" trait. But smart scientists have proved there's value in this. It's more than vanity; it's about enjoying the world around us and learning from each experience.

Paying attention to people has led to amazing discoveries. In 2015, Erin Smith, a fifteen-year-old from Lenexa, Kansas, watched a video featuring Michael J. Fox. The actor was diagnosed with Parkinson's disease in 1998. Something about the video nagged at Erin. His smile and his laughter seemed off. "It came off as really emotionally distant," she told reporters. She checked in with Parkinson's clinics to see if they'd noticed that too. Yup, they said. But the staff there gave her a nugget—similar expressions are found in patients, often years before a diagnosis has been made. *That sucks*, she thought. *The earlier you find it, the better your outcomes.*

Always a science nerd, she had an idea. In her mom's kitchen, she built a super-smart selfie camera that captured changes in facial expression over time. She called it FacePrint—the first objective measure of face changes over time. Her diagnostic algorithm has an 88 percent accuracy rate (compared to 81.6 percent for comparable but more expensive options). Michael J. Fox was so impressed that he funded her work. There are other baes like her; you probably know someone. Or you could be one.

Pree will start shipping the first consumer Nailbots in winter 2020. She's partnered with some big brands (she can't tell me the names yet, but she

said I'd recognize them!) to showcase a bunch of unique nail designs on the app. The fact that they've partnered with her adds extra validation to her business, but, even so, she knows that she could still fail; 90 percent of all startups do. She's sacrificed a lot to get here; she rarely dates and struggles to fit in workouts and time with friends.

"I can't predict the future. No one can," she says. "But I'm a firm believer; whether you're a politician or whether you're an entrepreneur, stay alive as long as possible. Learn from your surroundings. There's going to be a moment when maybe your message is going to resonate, because things shift." It would suck if that never happened for her, but she says that's okay; in the big picture, it's better to try to change the world then give up before you've started. She's already made a difference in so many lives, and this won't stop her.

CHAPTER 2

THE FOOD OF THE FUTURE

On January 22, 2018, inside an open-plan office, I and others crowded around a small trestle table manned by Kimberlie Le, twenty-two, a slim Asian girl dressed in a black and white striped top and blue jeans. She handed me a small plate with a small salmon fillet, two crackers, and three carrot sticks. Then she served the next person. And the next. In five minutes, all her samples were gone. She watched people eat her samples, noting how they bit, chewed, and swallowed every last bite. When they'd finished, she collected their plates with a wide smile. "It's really good," a man in jeans and a black hoodie told her. "It tastes just like salmon." She nodded her thanks. But while salmon and her sampler salmon are nutritionally similar in terms of protein, fat, and omega-3 fatty acids, her salmon was born from algae, instead of a cluster of eggs.

There's a bigger reason for Kimberlie's fake fish than pure novelty. At heart, she wants to change the world, one tasty bite at a time.

That's her answer to the bigger problem: why is the world she lives in so messed up?

It's no big secret that climate change is affecting the way people live. Oceans are rising, deserts are spreading, and polar ice caps are melting—resulting in tornadoes and floods that are ripping the world apart. Global warming has led to an increase in forest fires and millions of people are dying—each year—from the cumulative effects. Decades of people not giving two shakes about the environment has driven the Saudi gazelle to extinction, and the black rhino, Amur leopard, and Japanese sea lion are down to their last hundred or so members.

This is her inheritance—and it's rushing headfirst into an all-out global food and environmental crisis.

By 2050, the earth's population will have swelled to 10 billion people, up from 7.7 billion today, and if something doesn't change—fast—that means unprecedented levels of starvation, disease, and even doomsday-esque destruction.

That's such a big problem that Kimberlie can't quite visualize it, and she knows she can't fix everything on her own. But she can work on a piece of the problem.

One driver that's sending earth to its doom is the fish and meat market—as in animals; cows, pigs, salmon, chicken, etc., that are bred and fed for the dinner table. The energy involved in rearing all those

finger-licking chicken wings accounts for 25 percent of all greenhouse gases.

One way to halt that is by switching to plant-based food—that will take an enormous strain off the environment.

"We need sustainable and nutritious food," Kimberlie said. "Time is running out to reverse the effect of what people have done to the climate."

* * *

Growing up, fish was a big part of Kimberlie's diet. Born in Edmonton, Alberta, Canada, she was the eldest of four children. Her parents had fled there during the Vietnam war, seeking a safer, better life for themselves and their children. Her mom, Chi Le, was one of eleven siblings—most died in the war. In Vietnam, Chi had worked as a cook in restaurants, and she was the self-appointed chef of the family. Kimberlie ate fish with most of her meals: steamed, sautéed, fried, braised—pho noodles with fish sauce, Chả Cá Thăng Long (Vietnamese Turmeric Fish with Dill) Chả Cá, a grilled white fish and turmeric dish, fish cakes in banh mi, Ca Kho To (Caramelized Fish in Clay Pot), and fermented fish sauce to add flavor to each dish. She was taught to finish everything on her plate; her parents always provided, but they had little cash. She learned not to waste anything. But

she knew she was lucky, as many had people had it far worse than she did.

She loved her mom's food, but it marked her as different, as foreign. Edmonton was a friendly city, but only 1.5 percent were Vietnamese, and her skin color and fish-sauce condiments were unusual. But it also meant she was special, her parents said. They capitalized on their "foreignness," scraping together their paychecks until they had enough to open a restaurant. All the kids helped out; Kimberlie's been a server, a host, a dishwasher, a runner, a busser, and a sous chef.

In her free time, she loved snowboarding and competed with her high school team. But each season on the slopes, there was less and less snow in the mountains. "That placed climate issues at the top of my mind," she said. But she refused to get depressed by this. "It motivated me to make a difference—to make the world a better place to live," she said.

She graduated high school two years early and used the time to explore her interests; she was curious about so many things—music, arts, the environment, languages. She took a gap year in Shanghai, where she studied Mandarin and environmental engineering at Donghua University. Fish was a big staple here too; China's the number-one consumer

of fish in the world. This was a problem, worse than being in the mountains. In Shanghai, the air was thick and heavy with pollution from the coal-burning factories' production of cement and steel. People wore face masks when they ventured outside—they used these so much that you could buy them in rainbow colors and designer brands. Cancer rates were skyrocketing; 37 percent of the world's lung cancer patients were in China. You couldn't drink the water—97 percent of it was filled with toxins and pathogens. Fish were dying in hundreds and thousands from the polluted rivers.

Fish are dying across the world—climate change is having a dramatic effect on sea life.

It crystallized for Kimberlie when she went diving in the coral reefs off the coast of Taiwan. The water was blue and clear, and, beneath the surface, she marveled at the electric blue and neon orange fish darting around her legs. Some were striped like zebras, others had large luminous eyes that solemnly observed her—a real-life *Finding Nemo* seascape. She came back the following year, excited to venture below once more. This time her eyes widened from horror, not awe. The reef had shrunk, folded in on itself, the colors muted, and there were fewer fish. The vibrancy and magic were gone. "A year doesn't

seem like very long, but [climate] changes are happening that fast," she said.

There's a reason we are seeing this first in the oceans. The sea is a buffer for dry land; it's taken the hit from greenhouse gases to protect the earth. Oceans absorb 93 percent of the heat trapped by greenhouse gases. The rising temperature of the water is the reason for many of the fish deaths—scientists estimate the fish population has dropped up to 35 percent because of global warming and historic overfishing. That really sucks, because fish protein makes up around 17 to 70 percent of people's animal protein diet globally, according to a UN Food and Agriculture report. That's millions, maybe billions, of people going hungry.

Sure, some fish have benefited from the warming sea; the black sea bass population has grown 6 percent as the hot water killed off all their competitors. But this is still bad for them in the long term; when it gets hotter, they too will die. Climate winners will not be winners for long. "Fish are like Goldilocks, they don't like their water too hot or too cold," said Malin Pinsky, an environmental and biological science professor from Rutgers University.

The reduction of fish available has meant that fish farming—where schools of fish are bred in

captivity—is growing. But these schools spit carbon into the air.

In 2013, eighteen-year-old Kimberlie didn't know all of this. But she knew the oceans were important to her, and that she wanted to be part of the solution.

In September that year, she started her undergraduate degree at the University of California, Berkeley. She'd enrolled as a music major, hoping to work on her piano skills and take classes that caught her eye. She quickly realized that was a flawed plan. There were so many interesting classes on offer, and she didn't want to miss out. She didn't know what she didn't know. "I needed a multidisciplinary education," she told me; she needed a broad understanding of what was going on to attack environmental issues from "scientific, economic, legal, and social points of view."

So she changed her path, opting instead for a triple degree: a BA in Art and Legal Studies, a BS in Science, Society & Environment, and a Bachelor's of Science with a Concentration in Molecular Toxicology. Plus, minors in music and food systems in food science. "Everyone advised against my course load," she said. The college warned her that she'd be overextending herself and her friends suggested that she might not get the right college experience, but she was determined. "I wanted the fullest breadth and depth of understanding possible." That was the reason she'd picked Berkeley after all—she wanted to be somewhere where "people shared a love for hustle and getting things done," she said. "I wanted to meet people dissatisfied with the status quo."

She completed her course load in three years and used the remaining two to work on her own projects at the school's labs. She wasn't sure what the answer was—yet—so she experimented; she had petri dishes full of microbes and fungi to help her better understand natural ecosystems.

In 2017, she took a class called the Plant-Based Seafood Collider Workshop, at Berkeley's Sutardja Center for Entrepreneurship & Technology. The premise of the class was to cook up a faux fish alternative. Students paired up into teams—the idea was to compete with each other for the glory, and for the five thousand dollars given to the winners.

Kimberlie liked the concept. It seemed very logical to her. The alt-meat market had boomed, with Beyond Meat and the Impossible Burger making headlines regularly (even before they were served in Burger King and Dunkin' Donuts), so focusing on fish seemed like a smart solution. There was a fish shortage. The ocean couldn't handle more fishing. If she could find a plentiful natural resource and

turn that into fish, that would solve a lot of things, she thought.

"Growing up with little money and little food on the table...my personal goal is to ensure that food is a right and not a privilege for everyone," she said.

At Berkeley, Kimberlie joined forces with Joshua Nixon, a bioengineering major she knew from a prior project at the entrepreneurship center—they'd tried to design a ski helmet that reduced concussions. She felt very strongly about that; as a semi-pro snowboarder, she knew how easy it was to get injured. They'd won that contest, and she knew they were good as a team. "Let's be Team Dory," she suggested—she liked the idea of referencing the movie. After all, weren't they too venturing into the unknown looking for something special?

Food chemists and protein specialists gave guest lectures to help the students understand the components of flavor, texture, and smell that make fish such a valuable food source for billions of people. Developing something with the same level of protein as fish was fairly easy, but getting it to taste like fish was a bigger challenge. They wanted to synthesize a salmon burger, but it had to have the right kind of flakiness and texture. Their competitors experimented with plant proteins, but Kimberlie persuaded Josh that fungi were the obvious go-to.

She singled out *koji* in particular, as it has similar properties to meat; when cooked, it has a textured meat-like structure and an umami flavor. It's familiar to a lot of people, as it's used in miso soup and soy sauce. It's also a common ingredient in Vietnamese cooking. She enhanced their salmon with microalgae, which loaded their Frankenburger with omega-3 fatty acids. "Repurposing it from the lab to the kitchen to prototype on a sterile environment was the last step," she said.

But how to make it taste and feel like fish? She needed to figure out three things: flavor, smell, and texture. To do this, the pair created flavor profiles, testing and retesting different combinations of ingredients till they got the taste they wanted. The smell was a separate challenge—using gas chromatography to sort every substance in her product, she figured out which elements created the aroma she needed—and what she had to do to maintain it. "*Koji* fungus is high in protein and neutral-tasting to start—we combine natural flavors and other ingredients like plant-based fats to create seafood-like products." She found the class challenging; there was so much information and so many problems to address in the faux fish space. "We had to focus and work with the limited time

and resources we had to create first iterations of our products," she said.

They were awarded first place and given the five-thousand-dollar prize. They also got a surprising suggestion from one of the judges: "Have you thought about applying for IndieBio's next round?"

IndieBio's the premier San Francisco incubator for biotech startups. Teams that get accepted into their cohorts get $250,000 in seed funding, free lab space, materials, and four months to create a prototype for their demo day. Their application was approved. It validated her theory. "People really want great-tasting food that's good for them. And that's what we're here to deliver," she said.

She dropped out of Berkeley to participate, and she and Josh drew up official company documents; they called their company Terramino (*Finding Dory* felt too childish for a real business). Later, they rebranded as Prime Roots, as a nod to their fungus-based products. "Are you sure about this?" asked her mother. Kimberlie thought about it. "Yes, I really am," she said. "I'm confident I can execute, and I think it's really interesting. I really want to do this." "Then we support you," her parents said.

She worked long hours at the incubator. Every day, she encountered another hiccup—algae that

congealed, beakers that leaked, timers that didn't work. "Nothing derailing," she said, just the fun of having to do every single job. Her production efficiency grew 500 percent.

They left the incubator on a high. People raved about their salmon burgers, declaring them uncanny and delicious! Her name was getting out there—in June 2018 she received a $100,000 Thiel Fellowship.

Peter Thiel is a Silicon Valley legend, a billionaire best known for founding PayPal and for secretly backing wrestler Hulk Hogan in the Gawker Media lawsuit. In 2010, he announced the Thiel Fellowship: he'd give $100,000 to people age twenty-three and under who had a business idea they wanted to pursue. The caveat was that they had to be willing to drop out of school to take part.

When Kimberlie first heard about the fellowship, she thought it was dumb. "I loved school," she said. "It's the best way to learn as fast as possible." But looking at her Berkeley classmates, she started to see the point. Many of them were unhappy with their studies—they had so much potential, and it was being wasted. She'd already dropped out, so she qualified (they've also removed that requirement). Getting the fellowship is prestigious—she used the media attention to leverage $4.5 million in funding

from venture capital firms. "We found people who believed in the world we wanted to create," she said.

Having people believe in what she's doing gives her hope for the future. "America was built on the backs of immigrants," she said. "I think we are working toward a more diverse culture. Having a wider perspective is important. It's a white man's world—and [they're] wearing Patagonia fur."

Kimberlie's not the only #girlboss in the fake fish space.

Another pioneer is marine biologist Dominique Barnes. Dominique grew up among the neon lights of Las Vegas, her closest access to sea life in the aquariums at The Mirage and Golden Nugget casino. She loved visiting, pressing her freckled face against the glass, her brown hair pulled back in a ponytail, so she could get a closer look as the sharks and tropical fish swam by, blinking silently at the never-ending stream of gamblers.

Fascinated by sea life, she applied for a Master's in marine biodiversity and conservation at Scripps Institution of Oceanography at the University of California, San Diego. Being so close to the ocean, she saw animals die every day, choking in polluted waters, bleeding from injuries caused by careless

fishermen, or garroted by plastic rings from canned six-packs.

It was getting worse, year by year.

And no one seemed to be doing anything, not really. The people in power weren't doing much, especially the men. They might murmur that they cared, but nothing changed.

Typical. In general terms, the male of the species shoulders most of the blame for climate change. The patriarchy has held historical control over power, money, and land use, and, time and time again, they've made bad choices. Maybe women would also have made mistakes—but they never got the chance.

And now women are cleaning up the mess that men left behind; there's a historical precedent for women cleaning up after men, in kitchens, in households, and in the emotional labor of bringing down a system of sexism, racism, and inherited bias. Climate scientists aren't immune to this; among researchers, women are often dismissed and talked down to.

That sucks because women bear the brunt of this mess—80 percent of people displaced by climate disasters are women and girls, and they're fourteen times more likely to die during natural disasters.

Men also have a bigger carbon footprint than women, and overall, they care less; a 2019 study from Yale University found that men see caring about the world they live in as feminine. "Women on every metric you can think of are more environmentally friendly," said Ashley Johnson from Do the Green Thing, an environmental nonprofit. "They are more likely to recycle, they litter less, they are more likely to buy an electric car, they are more likely to vote for politicians who care about the environment."

Johnson believes the gender dynamic is to blame here. "Climate change is sexist," she said. "It simply magnifies the existing inequalities within our society. The research suggests that some men fear that green actions will brand them as feminine."

In San Diego, a mutual friend introduced Dominique to materials scientist Michelle Wolf. They clicked immediately, both sharing the same worldview about a sustainable food system and the horror of the depleting oceans. Together, they hatched a plan for a fish-free alternative, brewed up from plants and algae.

But what fish should they target first? Every pound of fish caught in the wild resulted in five pounds of other marine wildlife that was discarded, including dolphins and sharks. After a deep discussion, they chose shark fin as their first product—it was

considered a delicacy in some places, which had led to illegal shark hunting.

They named their imitation shark fin Smart Fin and made it by binding yeast with collagen to mimic the fin's texture. They got some pushback from conservationists who didn't want people encouraged to eat any kind of shark—even a fake one—as food; since 2013 it's even been banned from state banquets in China.

Frustrated, they pivoted, choosing shrimp as their next product. It's the most popular seafood in America, with one billion pounds consumed every year. And shrimp production, whether by farming or trawling, has some serious side effects. There is a high fatality rate for workers and a record of slave labor use on the trawlers. Shrimp farming is really bad for the environment, with around five square miles of mangrove trees felled to produce just two pounds of shrimp.

At UCSD, Dominique had run classes for school kids about the wonders of algae, and that made her think it could be a good base for faux shrimp; it was high in protein and could be adapted for multiple purposes. "Our shrimp isn't made in a lab. The process is more akin to making bread, and algae is our flour," Dominique explained. "We focused on

texture first." Their recipe is a secret, but seaweed and algae are big elements.

They got down to work, naming their company New Wave Foods to herald a new era of sustainable protein. "No one is looking at seafood this way, so we fit nicely with alternative meat," Dominique told reporters. They also added another talent to their board, Jennifer Kaehms, a biomedical engineering student at UCSD.

With a grant from IndieBio, they had a prototype four months later. Their shrimp was small and pink (they used red algae for color) and tasted springy and chewy and a little crunchy when I popped one in my mouth. "We're not reproducing shrimp cells," Dominique said. "We use a process that's similar to baking a loaf of bread." Their samples ran out in five minutes. Media interviews followed.

* * *

In her first few TV appearances, Dominique looked excited and nervous. On CGTN America, she stood next to her cofounder Michelle, both in black blazers (Dom's accessorized with a green coral-like necklace), their hands clasped behind their backs. Her grin looked forced as the correspondent popped a shrimp into his mouth. But then he asked her a question, and the tension lifted from her face. She

happily talked about the science, and the oceans—this she was comfortable with.

“Did you know that we’re kosher?” she told another host. It sounded like the beginning of a joke; three rabbis walk into a biotech incubator... But that really happened, and they’d really approved their shrimp for Jewish use. Dope.

By another stroke of luck, Google—which has made sustainability a company priority—approached them, offering the chance to stock their popcorn shrimp in its cafeteria rotations, feeding hundreds of hungry techies. “At Google, we have a lot of allergies and dietary restrictions,” Chef JP Reyes said. “It’s nice we can offer a shrimp alternative.” He combined their shrimp with panko and seasoning and put it on the grill for that slightly crispy texture. Google ordered ninety kilograms of their shrimp. Another early customer of New Wave Foods was the Monterey Bay Aquarium Cafe—it’s the first plant-based alternative food served on their menu.

“Disrupting Seafood, Not Oceans,” their slogan said (now they use the slogan “Plant-Based Deliciousness”). Of course, any kind of processing has some carbon footprint, but using algae minimizes this considerably. An average shrimp farm has a six-year lifespan before the water becomes so toxic that another mangrove clearing is needed.

They had some hiccups. Jennifer left the team to work on her own projects, and they scrambled to cover her role. Bye, Felicia.

They also battled the bias people had toward eating algae. "When I talk to people, usually they're like, 'What are you talking about? This is pond scum,' " Dominique said. She tried hard to change their minds. "Algae is a part of the foundation on which the ocean is built!" Sometimes she wondered if she'd ever get over that hurdle.

Then she got a lucky break. In 2018, Mary McGovern, a growth and brand executive with thirty years of experience in the food space joined their team as an advisor. She was so enchanted by their product that she then took on a CEO role, giving Dominique and Michelle time to focus on the product. Mary negotiated a deal with Tyson's Foods, who bought a 10 to 15 percent stake in their company in 2019. FYI, Tyson makes forty billion dollars a year in meat—by investing in faux fish, they're telling the girls they're doing something right.

They're currently stocked in two restaurants in San Francisco and one in New York, with more planned for 2021.

"Our oceans are overfished, and their temperatures are rising," said Larissa Zimberoff, a food and

technology reporter and author of *Technically Food: The Business of Plant-Based Meat and the Battle to Control What We Eat*. "This will affect our global ecosystem in unknown but potentially detrimental ways. However, with the increased demand for seafood, plant-based and cell-based analogues could finally help to divert fishing from the oceans."

She noted that cell-based seafood will "almost certainly be healthier" as it would be free of antibiotics, microplastics, and mercury. "Millennials and Gen Z are looking for food that's created by mission-based companies, is sustainable, and has a transparent supply chain," she said. "It's going to take time before widespread adoption happens."

Kimberlie and Dominique are doing their best to save the world, one bite at a time. But they're also business-savvy. The plant-based food space is blowing up—it made around $4.5 billion in 2019—and they're dominating the alt-fish part of the market. The global pandemic has only increased the growth of plant-based foods, as people become more conscious of their consumption. The Plant-Based Foods Association reported that sales of all plant-based products were up 90 percent from the previous year in April 2020.

Kimberlie's not slowing down. In fall of 2019, she relocated her office to Berkeley from San Leandro.

She's pleased with the move; they have great kitchens and a space she's turning into a restaurant. In 2020, her goal is to get Prime Roots products to more people. A number of products are in the works, including sausages, chicken tenders, protein bars, crab cakes, lobster, shrimp, and salmon burgers. In February 2020, they shipped a limited run of bacon. In the runup to launch, their products are not available to the public, so the only way to get a taste is at one of her private dinners (free to attend; sign up on her website).

Things are looking good, but she's not slowing down to celebrate.

"Let's keep going, because time is running out. And we only have one planet," she said. "All the first-world problems that the generation before us were pioneering, they don't really matter—if we don't have an earth, like, we don't have a place to live..."

To relax, sometimes she'll sip a complicated mocktail or a CBD-infused tea. "Myself and many Gen Zs are into beverages that don't mess up their mental development," she said—she thinks that's a big problem with millennials. To relax, she snuggles with her puppy, a part-German shepherd, part-pit bull, part-Chow rescue she named Chuy, after Chewbacca. "He's our chief cuddle officer," she wrote on Facebook.

Technically, twenty-four-year-old Kimberlie is a college dropout. But she's likely the most educated college dropout you'll ever encounter, with a 99.9 percent completion of three college degrees from UC Berkeley. "I just need to press a button, and I'll have the three degrees," she said. Her mom bugs her about that all the time, she said. She likes having an almost-degree in her back pocket; it gives her leeway to take more courses if something comes up.

For now, she's laser-focused on making sure people never have to go hungry like she did. That's what matters; fame is irrelevant. "I want to create something that touches millions or billions of people in a meaningful way—but I don't have to be *known* for that," she said. "If I can feed billions of people with sustainable protein—then I'll be happy."

CHAPTER 3

WANNA PIZZA ME? YOUR ROBOT FUTURE

In 2018, Vivian Chu, cofounder of Diligent Robotics, drew up a list of the most important features that Moxi, her hospital helping robot, needed to have. The key things were that Moxi would be safe, reliable, and cute. Hospitals don't normally factor cuteness levels into their operations, but then they don't have to think about how nurses and patients will react to a glossy white, five-foot-five, one-armed robot. When Moxi powers up, her LED lights glow and blink, and, when she's really happy, small hearts appear in the center of her eyes—a reaction to hearing people mention her name. "We wanted Moxi to be a teammate," said Vivian; that means nonthreatening and able to perform simple tasks like fetching linens and medications, ideally slotting seamlessly into the current system.

Inside the Texas Health Presbyterian Hospital in Galveston, she watched nervously as Moxi made her debut, rolling through the corridors. Would people be creeped out? Would they get angry? Would the nurses see Moxi as a resource or as an interloper?

At first, some of the nurses were suspicious of Moxi. Was this another high-tech tool they needed to learn

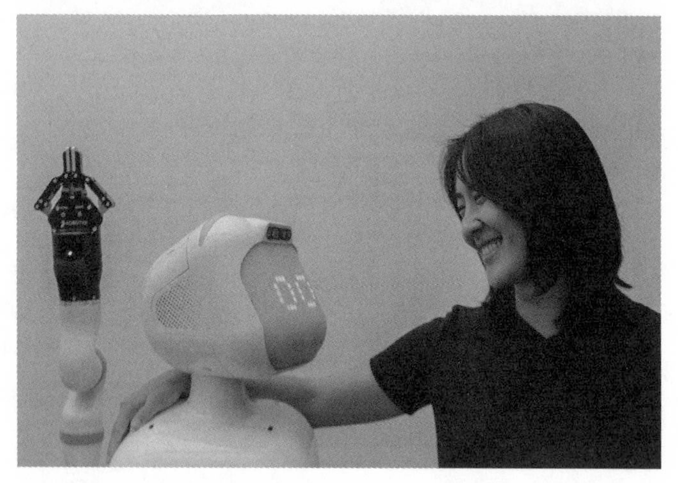

Photo Credit: Diligent Robotics

how to use? Would Moxi get in their way as they did their everyday work?

Vivian had planned for this. Moxi—preferred pronouns she/her/they/them—could only access the hospital hallways, and could not enter patients' rooms, for that exact reason. She'd also equipped Moxi with smart spatial sensors that enabled her to get out of people's way and had programmed her to react to people's inquiries by tilting her head to one side and emitting small beeps. On a daily basis, humans receive similar nonverbal cues from people, so mimicking them in Moxi would help everyone adapt more quickly. "That's one of the challenges of social robotics," said Vivian. "We have to figure out how people communicate with robots, and how the robots understand that and communicate back."

The nurses soon warmed up to Moxi. "Good morning, Moxi!" they said when she entered their wards. "Nice seeing you, Moxi!" Moxi was programmed to start work two hours after the nurses went on their shifts (so that she wouldn't get in their way), but she got some heat for that. "The nurses started saying, 'Oh, you're so lazy, Moxi, we've been working for hours!' " Vivian said. She adjusted Moxi's start time. Moxi quickly became one of the crew; nurses chatted to her on their shifts and took selfies with her on their

breaks. Vivian was relieved—she'd hoped Moxi would fit right in, but you never know, right?

Moxi fulfilled Vivian's cuteness brief, but entertainment and selfies were secondary to her real mission. Her cuteness was a means to an end. She'd been created to fill a real need. People nowadays are living a lot longer than ever before—which is awesome!—meaning that the world needs to be equipped to handle and care for an aging population.

* * *

Right now, there are massive shortages of trained people in industries ranging from medicine to physical therapy to teaching and even hospitality. With one in five Americans projected to be sixty-five years old and over by 2030, this is becoming a bigger and bigger problem. If nothing changes, by 2030, California will be short 45,000 nurses and Texas will be short by 15,000 nurses. Around 800,000 new nurses are needed by then—but there are nowhere near enough people to handle it. Reasons for this include low pay, on-the-job violence (due to staffing shortages), burnout, and a lack of investment in skills training.

Robotics is one way of solving this. Many people are scared by the idea of "robots taking our jobs" but the

reality is that we don't have enough people—right now—to fill those jobs.

The idea is that robot automation can fill the gap between supply and demand; they'd complement the existing services, not replace them entirely. It's more than a lack of nurses—it's also about preventing burnout from working. In nursing, in particular, it's women that bear the burden of overwork—91 percent of nurses are female, and burnout rates are around 30 to 40 percent. That means thirty-five or so out of every one hundred nurses leave the profession. Nurse burnout is bad for their health—and, indirectly, everyone's health.

* * *

That's where Vivian hoped she could play a part. Today's nurses have so many demands on their time that they only spend 37 percent of their working hours with patients. The rest of their time is spent on chores—which consist of everything from changing sheets to cleaning floors. If Moxi could help nurses with some of the boring nonmedical chores, they'd have more time to spend with the people they were caring for, she realized. "It's about giving them time to focus on what matters most," she said.

Working in healthcare is a new field for Vivian—medicine had never been part of her life plan.

Growing up in the San Francisco Bay Area, she was obsessed with video games and playing sports at school; she played tennis and basketball and dreamed about making the NBA one day. Sure, it was all men right now, but that would change, right? Every year. she looked at herself in the mirror and hoped she'd shoot up in height. It would be hard playing ball as a queer Asian woman, she knew, but if she was good enough, that shouldn't matter. But when she realized that five foot four was as tall as she was going to grow—the shortest NBA player ever was five foot three, but that had been in 1987—she figured that dream was dead. Her second favorite subjects were math and science, so that was probably a more realistic plan for her, she decided.

She went to UC Berkeley for her undergraduate degree in electrical engineering and computer science. She was going the science route, but she wasn't sure about what her specialization would be. Her subject was so broad! She signed up for Programming 101, but the class wasn't what she'd expected. Everything was so complicated. There was so much theory! "I could install Windows...but that was about it," she said. She struggled to complete the class, spending hours on homework where other students were done in an hour. She just couldn't connect the abstract numbers on her screen to

things in the real world. *Maybe this isn't for me*, she thought.

In her junior year, she signed up for an Introduction to Robotics course. In one of their first sessions, her professor brought a Roomba vacuum cleaner to class and tasked his students to design programs for it.

Vivian programmed her Roomba to scuttle up and down a ramp, using the sensor on its head to handle the elevation differences. "That was when I realized that programming is about actually making changes in the world, and that a robot can sense and then react," she said. Watching the robot respond to her commands in real life was eye-opening. Something inside her clicked. That's when Vivian started finding programming fun. "It was, okay, I can see how this can have real-world impact, I can see how this robot can affect and improve people's lives," she said. She spent hours in the lab, sometimes working for thirteen hours at a time, absorbed in her projects. "I forgot to eat," she admitted—she was so busy learning. During her summer breaks, she interned at tech hubs around the Bay Area, everywhere from IBM to Google X.

The robots she worked with were nothing like the robots of her childhood movies—a Roomba was

about as far from R2-D2 (her favorite) as you could get—but even so, she could visualize their potential.

After college, she applied to a bunch of graduate research programs that featured robotics labs as a key part of the course. Most programs were run by men, really talented men, with visionary ideas, but one program in particular stood out—the Socially Intelligent Machines Lab at the Georgia Institute of Technology, run by Professor Andrea L. Thomaz. "I do feel more comfortable in a female-led environment, but that wasn't a huge part of my decision," Vivian said; Andrea's work with social robotics fascinated her. "Her lab was amazing," she said. "I was like, this is the future of where robotics is going." She started in 2013.

In 2016, Andrea called her in for a meeting. "I've been thinking about starting a company," she told her. "Would you be interested in being my cofounder?" It was an easy yes for Vivian. She knew that they worked well together, and she respected Andrea's process and drive. And she knew that she brought a lot to the table; it would be a partnership of equals. "I'd taken business classes at Berkeley, and being from the Bay Area, startups and entrepreneurship are always in the back of my mind," she said. "The idea of taking technology into the real world was so interesting to me."

The shift from advisor to cofounder was interesting to navigate. Vivian had to keep reminding herself to speak up more and not instantly defer to Andrea's opinion. They discussed what industry would benefit from robots the most and settled on hospitals as the places with the most pressing issues. "The reason was the impact and industry need," she said. "Technology can reduce some of the challenges."

To learn more, they got a grant from the National Science Foundation that gave them the freedom to spend time in clinical settings. They spent the next four months shadowing nurses in local hospitals—over 150 hours—walking up and down the lemony, antiseptic corridors with a clipboard and a notebook. It hit Vivian hard. It was overwhelming to be in hospitals all the time. She'd never been seriously sick before, nor had any of her close friends or family.

Everywhere she turned, she saw and heard people crying and yelling in pain, and the antiseptic smell irritated her nose. She watched nurses juggle a hundred tasks, running up and down the hallways as they supplied linens, medications, and more to all their patients. The nurses were on their feet for twelve-hour shifts and longer, but her feet hurt after a few hours and, by the end of the day, her body and mind felt sluggish and drained. The nurses were incredible. But there seemed to be room for

improvement. "I saw all these pain points that I empathized with." For one, it seemed inefficient that the nurses had to go all the way to the supply room at one end of the hospital and then run back again every time they needed clean sheets or gauze.

"I wanted them not to be stressed and able to get supplies more easily," she said—this wasn't an aha moment, just an observation after hours of walking up and down the wards.

They got to work in their labs, wiring the circuitry and inputting the code. Vivian and Andrea's first Moxi was called Poli. It resembled a glossy orange and white bowling pin on wheels with no face and an arm poking out of its middle, a little like *Aliens*. Something didn't feel right. They tried again, this time adding a head and face and moving Moxi's arm to the side. Having one arm was due to the constraints they were under—robotic arms start at around $30,000, and they wanted proof of concept before they added a second. "There's a lot of tasks a robot can do with one arm," Vivian noted. Moxi's arm included a seven-foot extender, so she could reach every pesky little corner and shelf, and a gripper so she could pick up things of all sizes. They also designed her to access electronic health records, so when there was a change in patient care, Moxi could react; if a nurse scheduled blood work for patient A,

Moxi would head to that room so she could drop the sample at the lab. “There’s this kind of immediate connection that people have with something that has a face, with eyes, and that is the kind of connection that I envision people having with robot teammates,” Andrea wrote on the company blog. “We want this to be a trusted member of the team.”

Happy with their progress, they approached investors for funding. There was interest, but the pair were told they needed to be less reserved. “The bar is higher because we’re women,” Vivian said. “They told us to walk in more confidently—if we beat our chests like guys do, they’d be throwing money at us.” That wasn’t Vivian’s style—she’d rather underpromise and overdeliver than make airy statements she knew wouldn’t come true. But their determination paid off: True Ventures believed in them and invested $2.1 million in their company in early 2018. They were good to go.

In the spring of 2018, they rolled out a beta test of Moxi with a cohort of four hospitals. “Moxi gives nurses a third arm they can rely on,” said Aliya Aaron from AMR Healthcare Consulting. “Finding outside resources that care about nursing workflow and are working to create new efficiencies is the key to helping address nursing burnout.” At all times, Moxi was shadowed by a team member, so that

they could address any problems that came up. And answer the many, many questions. “There’s a need for more education about what robots can do,” Vivian said. “People aren’t aware of that.”

It didn’t always go according to plan. For one of their demos, they tried to show how Moxi could collect and dispose of soiled linens. Moxi rolled from hospital rooms all the way to the disposal room with the linens, where the head of infection control had to sign off on them. As this was a trial run to show how it would work on a real day, Vivian and Andrea put their jackets in the disposable bags, which Moxi carried from the ward to the disposal room. The inspector signed off on the bags and told them how impressed he was with Moxi’s delivery, making Vivian’s and Andrea’s hearts do happy flip-flops. Then it was back to work as usual, walking up and down the wards with their clipboards.

Later, when Andrea went to fetch their jackets, she realized they weren’t there. All the disposal bags were gone, taken by the trash removal truck, the nurses told them. Panicked, they ran to the parking lot, flagging down the truck at the very last minute. Andrea climbed into the truck bed, trying to ignore the smell, as she sifted through the many, many buckets full of bags. *Is this worth it?* she wondered, thinking maybe she should just chalk it up to bad

luck. In the very last bag on the truck, she found their jackets—and car keys, and wallets, and more. She took an extra-hot shower that night.

* * *

Vivian is focusing on the healthcare field, but there are many other ways robots can help out with work. Hospitals are just one example of how automation can improve things in the near future.

For Julia Collins, an entrepreneurial foodie, the most natural fit for robots was pizza. Julia's always loved pizza. Thin crust with mozzarella, deep-dish with olives and pepperoni, or Italian style with extra pepper—she loves them all. If she had to pick a favorite type of pizza, only one, she'd choose a classic Margherita—"Tomato sauce, cheese and a little bit of basil!" But making it isn't her jam—as a Gen X Black woman, staying in the kitchen is not in her life plan. That's where her concept for the Zume pizza robot comes in: a robot that can spread sauces, toss dough, and assemble pizzas to order.

Julia grew up in San Francisco and was infatuated with food from a young age—the many flavors of the city made it the perfect melting pot for an inquisitive, taste-hungry mind. Food made her feel safe and happy and connected—her grandmother's cheese grits exemplified all those feelings. But her family didn't want her to work in food. Her grandmother cried and left the room when little Julia told her that was her dream.

Maybe you should get your degree first, her parents suggested, so she went to Harvard University for her undergrad, majoring in biomedical engineering. They were so proud of her—it was a tough school to get into, especially as a minority applicant. In 2017, only 5.35 percent of Harvard University students were Black.

But it didn't sit right with her; she wasn't being true to her authentic self. After a couple of years working at a security firm, she decided enough was enough and headed west to the Stanford Graduate School of Business; she was ready to turn her love of food into a business. During her break, she interned at Shake Shack, right at the beginning of their brand when they were still a tiny company. Watching them grow gave her the push to go out on her own.

In 2010, she teamed up with two friends to open Mexicue in Manhattan, a Mexican BBQ food truck that served the lunch crowd. (It now has three brick-and-mortar stores.) She followed this with Cecil, an Afro-Asian fusion restaurant she opened in New York in 2014. It won best restaurant in the city—but she wanted more. When her friend Alex Garden

approached her about a startup idea that used robots, Julia jumped at the chance. The vision was to make nutritious and delicious affordable pizzas, she explained—technology could provide the link between fresh food and accessibility. She left New York for Silicon Valley, ready to start again.

It was hard to be new in town—and harder again because she was pregnant. Julia had meeting after meeting with investors and liked to inform them of her condition straightaway—it wasn't like she could hide it, anyway. She was normally the only Black person in the room and the only woman. You either loved her...or you didn't. Most did, admiring her chutzpah, and by mid-2016 she'd raised six million dollars in funding (by November 2018, this grew to $423 million) and had a working pizza-making robot. They equipped a delivery truck with fifty-six pizza ovens and a robotic assembly line that spread sauces, sliced dough, and more. A human employee stayed in the back of the truck with the robot to run quality control and add extra toppings if needed.

"The dream is not to get rid of jobs for humans, but to free up their time while the machines do the routine repetitive tasks that people don't enjoy," she told reporters. "Through automation, we were able to create better, safer jobs, eliminating dangerous tasks like sticking your hand in and out of an eight-

hundred-degree pizza oven and letting a robot do that. You preserve the job [of a cook], but you remove the tasks that are dirty, dangerous, and dull."

It was instantly cost-effective—their pizza was made in the back of Zume's delivery van, meaning it reached the customer fresh out of the oven—minus the overheads of a traditional restaurant. Julia also pioneered a new style of pizza boxes, made from sugarcane and plant-based fiber; it kept the pizza from getting soggy, and was 100 percent recyclable.

She sourced their ingredients from local sustainable farmers and made sure that her pizzas were lower in sugar, fat, and cholesterol than the other pizza delivery services. In late 2019, analysts valued Zume at $2.25 to $4 billion, but, by 2020, the Zume that Julia had built no longer existed: that January, the company cut more than 50 percent of its staff and repositioned itself as a sustainable packaging and delivery business. So long pizza, hello compostable pizza boxes.

Julia left Zume in November 2018 to work on her next startup, Planet FWD, which focuses on feeding the planet while cutting down on the harmful emissions produced in food manufacturing. "I believe in the power of private companies to make a change," she told Black entrepreneur website

AfroTech. "I'm trying to feed the world and save the planet. That's what gets me out of bed every day."

Both Julia and Vivian took the people-first approach with their robots—enhancing the work life of people and taking away the icky parts. Not replacing their positions.

In 2020, Vivian's focus is on expanding Moxi's reach. She wants Moxi to be an integral part of hospital ecosystems within the next five years and, in ten years, thinks Moxi might be commonplace in nursing homes. Diligent Robotics has now raised $15.8 million, Vivian and Andrea's company has grown from the two of them to thirteen full-time staff, and they're looking to double that.

It was—and is—hard to run a startup and have a fulfilling social life, Vivian said. But she's figured out a process. She likes to start work early, getting in around eight o'clock so she can get a lot done while the office is empty. At the beginning of Diligent Robotics, she worked ten- to twelve-hour days, but now, unless it's an emergency, she leaves by seven o'clock at the latest. "It's hard to have that balance. You have to be conscious and build in a timeline," she explained. But prioritizing her free time—whether it's an hour of running or an hour of playing *Mass Effect* on her PC—makes her feel happier and healthier, which shows in her work. She's inspired by

Andrea, who makes a point of "forced disconnection" by leaving the office by five thirty every day to put her kids to bed.

Moxi's effect on people continues to surprise Vivian. For one, Moxi gets a ton of fan mail. Most of it is handwritten, often in colored pencils or crayons, from kids at the children's hospital who want Moxi to know that she brightened their day. "It will say things like 'Thank you Moxi, I love you!' or 'I'm so glad to be here,' " she said.

She's hopeful that Moxi will inspire a new generation of female roboticists. She attends lots of maker days and outreach events for girls in tech to get them interested. "They're always surprised to hear that you could make this a career, that robots are so developed nowadays," she said. "It's a lot of fun to talk about my experiences and show them that it's really fun to work on robots, and that it's totally a career path that they can do."

CHAPTER 4

THE CLIMATE QUEENS

In spring 2019, Etosha Cave, the founder of Opus 12, a CO_2 conversion startup, watched impatiently as her dishwasher-sized prototype was unloaded from the delivery truck at her lab in Berkeley, California. She checked the time on her phone. She had sixty minutes left before Bill Gates, the billionaire cofounder of Microsoft arrived, and she was nowhere near ready. She wanted him to see the reactor she'd worked on with her team; it sucked carbon dioxide emissions from the air and converted it into plastics, fuels, and other materials. She hoped Gates would be impressed with what they'd accomplished, as at scale, their machine could offset a third of global CO_2 emissions. But would they be ready in time?

"We were scrambling!" she said. Gates was there to visit the many companies at her startup incubator as part of the Netflix series he was filming. It would be so awesome if he saw their work! The next hour was hectic, but, by the time he arrived, they had it humming nicely and were able to present it to him. "At scale, our reactor has the CO_2 conversion power of 37,000 trees packed into the size of a suitcase," she said. Bill Gates approved of their project. "You might

be one of the first applications that fits that niche," he told her.

FYI, Etosha makes a one-second appearance in the Netflix documentary Inside Bill's Brain: Decoding Bill Gates, episode 3, at 10:13.

The world needs solutions like Etosha's reactor and needs them now. Climate change is not going away on its own. It's a global problem. The biggest contributor is the huge effect that our rising carbon emissions have on the environment.

Quick primer: A number of gases create the earth's greenhouse effect, including methane, nitrous oxide, and ozone. The predominant one is carbon dioxide, which makes up around 75 percent of all greenhouse gases. Greenhouse gases sound like a bad thing, but they actually keep the earth livable for all of us. They absorb solar energy and act as a buffer between the sun and the earth, letting in heat, but not too much heat, so we can all survive. That's why it's called the greenhouse effect, or as I like to think of it: planetary sunscreen. The problem is that by burning fossil fuels in larger and larger quantities, our atmosphere becomes overloaded, and that's rubbing away some of our planetary SPF.

This carbon overload is directly related to human activity, and the root cause of the increase in natural disasters. Too much carbon dioxide is the cause of all the extreme weather we've had lately—hurricanes, tornadoes, and floods. It's warmed up the seas and sparked wildfires in the forests. Right now, CO_2 levels in the atmosphere are the highest recorded, since, well, ever. In 2019, the Global Carbon Project reported 43.1 billion metric tons of CO_2 in the atmosphere, the highest CO_2 rates in recorded history.

We have only ourselves—and our parents and grandparents and great-grandparents—to blame for this. Human activities like coal burning and deforestation release *sixty times* as much carbon into the atmosphere as all the world's volcanoes.

Climate change is real (duh!) and it sucks. We're already seeing real-world consequences play out in the animal kingdom. In 2019, over two hundred reindeer in Norway died from starvation, as the changing temperatures had killed off the plants they live on. In 2017, 18,000 penguin chicks died in Antarctica, and, in Asia, the reduction in rainfall and increasingly hotter habitats has significantly decreased the number of babies born to Asian elephants.

It hurts people, too. Around the world, one in thirteen humans suffers from asthma, and there's

been a sharp spike in insect-borne illnesses, thanks to animals leaving their habitats due to extreme weather conditions. In 2018, malaria, killed 405,000 people, and 390 million people were infected with dengue fever in 2018, up from 1.7 million in 2010. If temperatures keep rising, scientists at the World Health Organization predict that five or six billion people will contract this in 2080.

* * *

Numbers like this are scary. "There's an enhanced sense of urgency," says Etosha. It's something she worries about most days. Many people are working on solutions—the Paris Climate Agreement of 2015, for example—but Etosha likes to have a more immediate effect on everyday life.

Most people look at the fumes that factories spit into the air and feel depressed and worried for the future. But when she looks at them, she sees hope—and opportunity. If she can suck them out of the factory, it could make a major difference to life as we know it. CO_2 emissions are hugely problematic, but she envisions a world where they're useful. Remember, carbon is a key ingredient in a lot of products that humans purchase—it's used to produce everything from airplanes to cars and fertilizer.

For her, it's personal. She grew up in Crestmont Park, a low-income, mostly Black suburb of Houston, situated half a mile from an abandoned oil and gas waste site. Her mom worked as an elementary school teacher, with a focus on science, and her father worked as a bus driver, graduating to quality control work in transit and construction. Etosha and her brothers (she's the middle child) avoided playing near the site, but it was always present: an ugly monument to Texas's overuse of fossil fuels. Over time, waste from the refinery leaked into the soil and water and was linked to rare forms of cancer and illness in her neighborhood. "In Texas, you hear a lot about oil and gas and its role in society," she said. "It really affected me—I always felt like we could do something better with the waste."

Houston's well known for its role in the space program, and Etosha dreamed of being an astronaut when she grew up—maybe she could be the fourth Black woman in space! But her brothers were interested in playing, not space, and her focus shifted. "[They] wouldn't play with me unless we were building cities or playing Teenage Turtles," she said.

Her high school was a magnet school for engineering; magnet schools are public schools that have a special curricular focus, such as STEM

or vocational training. Etosha's school regularly hosted talks by people working in Houston's major industries: oil and gas, wind, and solar. Texas, then and now, has the most wind turbines of all US states, and Etosha learned about the benefits of renewable energy, as well as oil and gas. "Texas isn't known to be a green state the way that California is, but it will surprise you," she said.

Learning about earth-friendly ways to power things was "a catalyst to working in the renewable energy space," she said. In school, she gravitated toward science and math and was part of the National Society of Black Engineers. "I was a science geek." She got straight As, but she was shy about her grades: sometimes she hid in the bathroom when report cards were passed out. "I thought my friends would make fun of me, or something," she said. She often felt lonely—she didn't seem to fit in any particular box, and it was hard to make friends. "I wish I could go back and tell myself not to worry about that and to just have fun," she said.

The more she learned about science and the world, the more she worried about the environment. It felt like every summer, bigger and bigger hurricanes ripped through Houston, leaving people homeless and floundering and having to rebuild their lives. They got worse every year. It frightened her. "We

were flooding every two years—then it was every year," she said. "I saw this acceleration of climate change effects."

At a conference for the National Society of Black Engineers, she met a recruiter from MIT. "He made MIT sound like this heavenly place where people do these really cool projects," she said. "I had my heart set on going." She could work out the details when she got there. But during her senior year of high school, a new university called Franklin W. Olin College of Engineering opened in Massachusetts.

It sounded very different from the other colleges she'd looked at—including her beloved MIT.

Franklin W. Olin's big focus was on engineering and entrepreneurship. "It was a startup college," she said. She liked the sound of their program and thought it would be cool to be part of their first-ever undergraduate class. "I could shape how the school would form!" She signed up. Classes were small—there were around seventy-five students in her year—but she got a lot of personal attention. She was the only Black student. It made her feel anxious, like people were watching how she performed. Growing up, her parents had said that "you have to be twice as good to be seen as the same," and she internalized that pressure.

She also spent a lot of time thinking about energy problems, and what she could do to help. In 2006, after she graduated, she interned at McMurdo Station in the Antarctic and helped test a laser diode for NASA. She lived there for the summer. It was awe-inspiring to see nature at its rawest, framed by shelves of ice, active volcanoes, and breathtakingly blue water.

From here, she headed to Stanford University for her Master's and PhD in engineering. Doctoral students are allowed to select the labs they want to work in, and she was intrigued by the work happening at Professor Tom Jaramillo's lab. His focus was on converting CO_2 into other useful gases by adding electricity, water, and proprietary metal catalysts. "I gravitated toward it because he was doing something with waste," she said. "I liked that it could have a big impact on climate change." She discovered that their lab built off experiments conducted by Japanese researchers in the 1980s. Their key finding was that copper had the ability to convert CO_2 into methane, one of the main ingredients in natural gas. "You can make synthetic natural gas out of carbon dioxide," Etosha explained. "That's *huge*!"

For the next three years, she refined the process. It was exciting to be working on a solution to the

global emissions problem. But, as her graduation date got closer, she felt restless. She didn't want to stay in academia her whole life. She wanted to take this technology out of the lab and into the real world! She approached Kendra Kuhl, a student who worked in her lab, and Nicholas Flanders, an electrochemical engineering student she'd met at a clean tech networking event, about her idea. Wouldn't it be great to transform harmful CO_2 emissions into something useful? This wasn't wishful thinking, she emphasized; she had the science to back up her claims. They loved her pitch, and the three teamed up, launching Opus 12 in 2015.

First, they needed a place to work from and some money to get it in motion. They got a $75,000 grant from the TomKat Center for Sustainable Energy at Stanford University and were accepted into the Cyclotron Road accelerator at Lawrence Berkeley Labs, a two-year program that helps environmental innovators turn their technologies into companies. This gave them free lab space and funding. That was just enough to get them started, but not enough to live on when split three ways. And Etosha was impatient to grow their team—she could do so much more if she had more people.

She went through their finances over and over but didn't see any way to make it work. Her biggest cost

was rent. But she had to pay rent, right? This sparked an idea. She handed in her notice to her landlord and moved out...into her car, a purple Honda Fit. For over three months, she showered at the gym or at a friend's house before settling into her car for the night. She parked it in the hills near the lab, so it was always quiet. "It was a nice place to camp," she said.

The team got down to the business of recreating photosynthesis at warp speed. "You take carbon dioxide in a plant, and you make sugars. Instead of sugar, we're making a range of different compounds," Etosha's cofounder, Kendra Kuhl, told the media. Their reactor let them break apart CO_2 and water molecules, and reform them into new molecules. At Stanford, she transformed CO_2 into sixteen molecules, but here they've focused on four so far, including synthetic gas, methane, and ethylene, which is used to make plastics, jet fuel, and diesel fuel. "We can create a revenue stream out of CO_2 instead of throwing it into the air," Cave said. "Our whole purpose is to take CO_2 and make valuable things out of it."

It sounds like magic, but it's science at its best, able to pull stuff out of the air and turn it into something useful. Their work convinced some of the biggest names in energy, including energy bigwigs Southern

California Gas who invested in their business. Etosha was back in a real bed again!

Today they've raised almost twenty million dollars, and their investors include NASA, Shell, and the US Department of Energy. NASA's interest stemmed from the fact that 95 percent of Mars' atmosphere is CO_2. If they could pluck carbon dioxide out of the air and make things with it, that would reduce their shuttle weight. Etosha's space dreams don't look so far away anymore. "The idea is that you can send astronauts to Mars without every single thing they'd need when they get there," she said. "With our reactor, one day they could make plastic out of the atmosphere."

* * *

Etosha's not the only activist in the climate cleaning space. Today, many, many people care very deeply about the world, some from a very young age. "2020 is the tipping point for Generation Alpha," said Laura Macdonald, EVP and Head of Consumer North America at Hotwire, a global communications agency. "This generation is so much more conscious of climate change."

Generation Alpha is roughly the generation of kids born after 2010—they'll be the ones who'll shape what the world looks like in twenty years. They're the

activist generation, explained Laura Macdonald. "The climate crisis is what drives them."

In 2019, she surveyed interviews from 1,001 Generation Alpha, kids seven to nine years old, and noted that technology had a big influence on their activism. "Thanks to Alexa and Google, kids can access everything," she said. "They're not growing up in a bubble like their parents did. They're growing up in an incredibly diverse world, and they're aware of the bigger issues in the world around them."

Not only do they care a lot, but they care a lot more than the adults in their lives. Laura's report found that 38 percent of Gen Alpha said recycling was very important to them, compared to 22 percent of millennials. She also noted that 95 percent of Generation Alpha are already climate crusaders, and 90 percent are very concerned about the environment.

"More parents are talking to their kids about it and treating them more like adults," she said. Gen Alphas' opinions matter for the future—and, surprisingly, for the present, Laura reported. They might not be able to vote, but what they say carries a lot of weight at home. "We found that 25 percent of Gen Alpha parents got their kids' opinion before making a purchase," she said. That's purchases of everything—toys *and* television sets.

Other startups are also invested in the clean air space. But it's a hard area to work in. Clean tech isn't considered cool by investors, as it takes a long time to materialize. Investment in this space dropped 30 percent between 2011 and 2016, from $7.5 billion to $5.24 billion.

* * *

In San Francisco, Davida Herzl, the CEO of Aclima, has been working on mapping air quality since 2008. "I started the company because I was frustrated that the climate change conversation was about abstract timelines and scales," she said. A lot of people were talking about the problem, but she didn't see anyone actively doing something to change it. "We needed tools to help create a more balanced existence with natural resources."

This is something she'd thought about for a long time. Growing up, her father ran a number of businesses, many specializing in factory optimization. He often brought her along on his trips to suppliers and factories. "It was a really visceral experience to see the source of those emissions and how they affected local communities and the environment," she said.

In the big picture, everyone's hurt by polluted air. The World Health Organization reports that

air pollution-linked illness is the biggest health epidemic of our generation, with 92 percent of the global population breathing unhealthy air. That sucks. And the people with the worst air pollution tend to be low-income communities of color.

It starts with knowing just what's in the air around you. "We take 20,000 breaths a day, but we know very little about what we're breathing," said Davida.

Most US cities and states use monitoring stations to map air quality, but they map large areas. No one looked at pollution from a block-by-block perspective. Davida thought that was a huge oversight. "Air quality varies block by block," she said. "We have global-scale data; what we're missing is local data."

At Aclima, her team—which is 50 percent women and BIPOC—worked on building smart sensors, a.k.a. "Wearables for the planet." They track everything from air pollutants to greenhouse gases. She started small, with static indoor sensors that measured office environments and schools. "They're designed to provide meaningful data about the places where people spend their days," she said. In Los Angeles schools, Aclima's sensors highlighted which districts had the worst pollution levels. The schools then installed better air filtration. "We can't manage what we can't measure!"

In 2014, Aclima trialed their mobile sensors in Denver, Colorado, affixed to three Google Street View car windows. The cars mapped urban air quality and traveled 750 miles in a month, mapping all the while. In total, they collected 150 million data points. She was thrilled with the results—they now knew exactly where the smog and ozone problems were in the city. Having block-by-block data created a helpful picture of where the persistent levels of pollution and emissions are coming from, which helped the city understand whether it's a one-off or an ongoing problem. They can then use this data to take action against climate offenders, through changes to their licensing agreements and increased funding in specific communities. "They literally could not see this before," said Davida.

In 2020, in partnership with Google Earth Outreach and Google Maps, Aclima outfitted a number of Google Street View cars with their sensors and sent them to map different regions of California. And that's just her starting point. "We're beginning with air, but we see a future where our sensor networks expand to include things like water quality."

* * *

Etosha and Davida have a way to go till they reach their goals, but they're on the right track. In the long

term, Etosha hopes to see her machines utilized in every factory in the world. "Instead of emitting into the air, you could just convert it using our system," she said. "That's the larger vision." It's a big proposition, especially as she needs to convince owners of factories that send their waste into the air for free to pay for a conversion service. Then there's the fact that her CO_2 conversion process uses a lot of nonrenewable electricity and water, which is also problematic. But that's temporary, she said. "We see a world where renewable electricity will be the cheapest form of electricity," she said; her job is to perfect the science part, while climate activists work on the rest.

Etosha hopes the earth is going to be okay in the end, but she's not 100 percent sure. "I'm hopeful that we still have time to reverse some of the effects of climate change," she said. "But it's happening so fast that it could be too late." But with her NASA projects and related work underway, she's getting us one step closer to space as a backup.

CHAPTER 5

REPROGRAMMING PRISON

In early 2019, Clementine Jacoby, a red-haired software engineer from San Francisco, trailed Leann Bertsch, the director of Corrections and Rehabilitation for North Dakota, as she went about her day. She watched her analyze reports from her officers, talk through problems with her staff, and field calls from all sorts of governmental organizations. But, unlike many of the people who work in the corrections field, Bertsch stood out as a visionary—open to new approaches and eager to bring the jail system into the twenty-first century. Around 1,300 people are jailed in the state—the fourth-lowest percentage of state population in all of the US. But for some, their experience is unique to the US.

Most US prisons keep their inmates indoors, shuttling them through a series of events—lunch, lessons, and lockdown. But inmates in the North Dakota system are called residents instead of inmates and live in dorms instead of cell blocks, and there's not a hint of neon orange anywhere. Officers address them by their first names, and they're given keys to their rooms. "Leann's trying to bring more

of the Scandinavian approach to incarceration to the United States," said Clementine; this type of setup is common in Norway, which has one of the lowest rates of crime and recidivism (committing another crime after release from prison) in the world. "You can't rehabilitate people if you treat them inhumanely," Leann told the media.

Clementine admired what she was doing, but it was clear to her that they weren't quite there yet. "It's tough to appreciate how hard parole officers' jobs are until you shadow them. It's not easy!" she said. "You're supervising caseloads of sometimes over a hundred people, all with different needs." Most of the officers and staff didn't have up-to-date information about what was working and what wasn't, and while some researchers were looking into this, they'd release a report once or less a year—it just wasn't very actionable. "They just didn't have a good sense of what was working or a good way of setting goals to hit them," she said. That's where she came in.

It's no secret that the criminal justice system in the US is messed up. It's biased toward people of color; they're disproportionately charged and sentenced compared to white people, with sentences for the same crimes 19.1 percent longer.

It all runs on a model of punishment, not rehabilitation—sentencing the mentally ill to prison

time due to lack of sympathy and understanding of extenuating circumstances. Problems include solitary confinement, inhumane treatment, and forcing pregnant women to give birth under armed guard with their legs shackled—remember, 80 percent of women serving time right now are there for nonviolent crimes. More and more, this system is hurting women—the number of women behind bars has risen more than 700 percent since 1980.

Today, the United States accounts for around 4.4 percent of the world's population, but 22 percent of the world's prisoners—more than any other country in the world. That's not the kind of society that screams fairness and justice—you know, the tenets of the US Constitution. If this doesn't change, we're going to end up living in an incarceration nation, with personal rights only relevant to people with the right skin color or the right bank balance.

There are many reasons for the criminal justice problem we're facing, including old, racially targeted laws, lack of treatment options and psychologists for people struggling with anxiety and addiction, and antiquated systems and processes that don't account for individual cases.

There are lots of smart and conscientious people working on these problems. The Last Mile offers coding bootcamps for men, women, and kids

behind bars, giving them a pathway to employment when they're released, and, in Texas, the Girl Scouts' Behind Bars program helps mothers and daughters connect and maintain family relationships.

But all these schemes and programs don't communicate well with each other. For example, a judge in California wouldn't know—or have heard of—a Texas pilot program that sentenced people to work a specific number of hours with animals instead of serving jail time, the idea being to stop them reoffending. (And it did! Re-offenses dropped by 70 percent.)

That's messed up, thought Clementine. *How do you know what's working if you don't talk to each other?*

Another thing that makes it hard for people to get back on their feet: technical revocations. These can happen when someone's released on parole and breaks a rule. A rule, remember, *not a law*—the rule infraction could be anything from having the wrong tags on their car to missing an appointment (even if they didn't have gas money). Break the rule, and they get sent straight back to the slammer. 25 percent of people in prison right now—that's millions of people—are there because of technical revocations. "That doesn't help anybody and just leads to more pain," Clementine said.

That's one of the reasons Clementine cofounded the nonprofit Recidiviz in 2018, giving up her fancy engineering job at Google—and all the free cake, coffee, and swag that comes with that gig—to work on solving this real-world problem. Recidiviz's ethos is about making information accessible and using that to change behavior—at an individual and governmental level. Without data, you don't know how effective things are, or aren't... You can see how Google helped set her up for this role!

For Clementine, it's personal. When she was five, her twenty-one-year-old uncle on her mother's side was arrested. He was sentenced to twenty-two years at an Idaho prison. She remembers her mom crying about it—everyone was so sad. "It was really traumatic for my whole family," she said. Her mom visited as much as she could—it was an eight-hour drive from their Utah home to Idaho—but Clementine wasn't allowed to visit him, as only immediate family members were admitted; she wasn't *his* kid. Getting many visitors can really help people emotionally cope with incarceration, but restrictions like this are common in every state.

It was a constant topic of conversation—the whole family lived a middle-class lifestyle, and he was the only one who'd ever been locked up. He was forty-three when he was released; he'd never used an

iPhone or watched YouTube or drunk a Frappuccino. He'd been locked up her whole childhood. Her whole family was excited—they threw a giant family reunion, with balloons and cakes and all their cousins.

But her uncle didn't know how to handle this strange new world. There were so many ways to recycle. People used terms he hadn't heard of. Everything was paperless. A few months later, he forgot to pay a fine, which violated his parole. He went straight back to prison. Her mom was devastated, but Clementine was frustrated. This would have been so simple to prevent. It wasn't fair—not to her mom, her uncle, or the overcrowded, debt-ridden prison system.

This cycle repeated over and over with her uncle, and the family shifted from hopeful to resigned every time he was paroled. Clementine knew she was lucky to get that much time with him; white middle-class men weren't as overpoliced as minority communities. His struggle solidified her interest in the criminal justice system.[1]

"Realizing how easy it was to go back to prison once you'd been in—and how hard it is to get out—made

---

1 A few details have been changed to protect the family's privacy.

me want to work in criminal justice reform," she said. "I wanted to figure out why it was so hard."

Clementine didn't know how everything worked, but she knew that it wasn't working.

She just didn't know how she fit in yet. In high school, she gravitated toward the arts, loved dancing and writing, and thought she might go into journalism. She applied to Stanford University for her undergrad—"Why not go to college at a country club?"—and signed up for an Intro to Computer Science class in her first quarter to fulfill a general requirement.

It blew her mind. There were so many parallels to writing in CS, like whittling down what you were doing to its most basic components. "The logic chains were fascinating!" She went into another class, then another, until she was on track to be a software engineer, almost by accident.

But she was still drawn to the arts, and in her second year at Stanford, she enrolled in a dance class called Dance in Prisons. On Wednesdays, her professor covered criminal justice research and ran an open discussion about the day's readings.

"It was foundational for me—I'd had this experience of my uncle, but this was the first time I'd studied the subject in any kind of more holistic sense,"

Clementine said. On Fridays, the class went to a juvenile facility and taught dance to the imprisoned teenagers. The professor gave them a list of rules to follow; they had to wear loose-fitting clothing, like sweatpants, baggy jeans, and loose shirts, and no red or blue colors—those might look like gang colors.

It was weird being behind the fence for the first time. "We were teaching dance, but we couldn't do any partner work," Clementine said—no touching was allowed at all, whether in correcting students' form or in tango. "That's why we mainly did line dancing or hip-hop, things where you're not actually required to interact." It's tough teaching dance when your body is hidden in baggy clothes, but the teenagers were so happy they were there that they made it work. One week, Clementine led the class—she chose her own music, but it had to be approved by the prison first. It was eerie seeing it from that side of the wall. "It was weird seeing how few decisions the inmates have to make," she said. "It's dictated when they wake up, when they exercise, when they go to sleep, and when they eat."

"Lots of the kids reminded me of Stanford students—they were very entrepreneurial."

She'd known about this, theoretically, but seeing it in person really hit home. It made it harder that they were teenagers—just a tiny bit younger than

her at the time. "When they'd normally be making a ton of decisions, good and bad, and learning from them, they just sort of missed that period of cognitive development."

Upon graduation, she was hired by Google as a product manager. "It was like all the fun parts of engineering, and none of the bad ones," she said—she liked the teamwork, the design work, the partnership of building something together and managing the working parts. "Google's amazing at giving a lot of responsibility to young people—no other industry hands the reins over in quite the same way to people just out of college."

After three years, she started thinking about criminal justice again. She went to talks and workshops and volunteered her time.

One thing that Google's known for is the 80/20 rule, the concept that employees are free to spend 80 percent of their time on work projects, and 20 percent of their time on their own projects. "You talk to your manager and you're broadly pre-approved," she said. This gave her the perfect opportunity to start figuring out her role in the justice space. She teamed up with other Googlers who were passionate about it, and they started going to meetings and workshops.

But no matter what meeting she went to or what paper she read on the topic, she kept finding one recurring problem: information. “There was this fundamental data gap that prevented us from understanding,” she said. Everyone, from both the right and the left, was really focused on incarcerating only the people who were a danger to society—but no one was making much progress. The place that was meant to collect this information, the US Bureau of Justice Statistics, had slowed down their reporting, and they weren’t getting all the data they needed to begin with.

Then she read the results of a two-day think tank summit in DC. It had been convened to chart the path forward in criminal justice reform. “They came out saying we lack the data to know what’s working, what the next steps are, where to invest, and how to benchmark progress.”

She started small—or at least, she thought it was small—by building an open-source code base to link data about different correctional systems together, including parole, rehabilitation programs, and even county-level data. “We were aggregating more and more data over time and building this historical and real-time view of the data.”

She thought this would help families and advocates understand more about how the sausage gets

made, to quote *Hamilton*. The reality shocked her. It wasn't just people outside the system who wanted access but *the people working in the criminal justice agencies themselves*. "There's a bazillion really, really high-impact questions and they didn't have the data!"

One of the states she spoke to told her that they ran ninety programs to prevent reoffending—but they only had the capability to assess three of them a year. "So, in thirty years, we'll know what was working in 2019," she said.

The people making sentencing decisions didn't have comprehensive sentencing data. That was her tipping point and the beginning of Recidiviz. Now, she could see what she had to do next. For the next five months, she did double duty—waking up at five o'clock to work on Recidiviz till ten in the morning, then working at Google from ten o'clock till eight in the evening and Recidiviz from eight till two in the morning. People who believed in her plans gave her some investment money, which she used to hire four full-time engineers to pick up the slack. "It was a crazy time!"

She was accepted into the Y Combinator summer class of 2019 (Y Combinator invests in early-stage startups) and resigned from Google to focus on it full-time. Y Combinator companies tend to do

really, *really* well—they were the first to back Reddit, Twitch, Airbnb, and Dropbox. She thought she'd been working hard, but she was wrong. "It was like being on a moving conveyor belt; we were focused on rapid iteration and growth!" Technology moves fast, but could the states keep up?

"We found ways to move the states we worked with along much faster," she said. By the end of the three-month program, they'd grown from just North Dakota to Idaho, Kentucky, Missouri, and Pennsylvania, with interest from four more states. "We had to be sensitive to the difference between government technology and consumer technology," she said. The idea grew and expanded into the assessment of the long-term policy results from various justice systems. "We're reaching into legacy criminal justice data silos and linking them to help policymakers and practitioners see the outcomes."

"None of this is rocket science at all from a private-sector perspective, but it didn't exist in criminal justice." She was worried the data would be so "crusty" that it would be unusable. "We spent five months proving it could be done!" Next, she turned her focus to prevention. "How can we stop more people going to prison?"

Take her uncle's revocation for an unpaid fine. At the state level, he was just a number in the revocation

column. They don't know anything more about him. Was he taking medications that day? Had he eaten? Was he supported? Having information like that about who gets revoked enables the state to plan ways to stop that from happening. If people are revoked for mental health behavior, they could create more programming to help with that, for example.

But Clementine wants to go deeper than that. Sure, you see that you need more mental health solutions. But states are dealing with limited budgets. "Right now, it feels very scattershot," she said—people don't know what works or doesn't. So why not select the "program that performs best for eighteen- to twenty-four-year-olds"—if that's the demographic the data highlights?

It hasn't been easy working as a woman in this field—especially when she looks like a cross between Amy Adams and Galadriel, with curly red hair that falls in waves down her back, with a penchant for wearing braids, pink jeans, and performing aerial gymnastics. "The two parts of my life are very white-male-dominated—technology and criminal justice," she said. "It's been challenging, but my whole career has been in a male-dominated field, so I've gotten used to it."

* * *

More women are making a difference in the justice space.

In 2017, Phaedra Ellis-Lamkins and Diana Frappier joined up to develop Promise, an app that uses computer science to identify—and eventually, help—the 8.4 million people a year who get stuck in jail, prior to sentencing (so presumed innocent), who can't afford bail. "It's un-American and it's unjust," Phaedra told a *Forbes* reporter. "A person may have taken a plea to a charge they were not guilty of to get out of jail because they couldn't afford to pay for their release and needed to get home to their kids or job."

Phaedra and her sister grew up in Suisun City, California, a town with less than 30,000 people. They were raised by a single mom who struggled to put food on the table. When her mom got a full-time union job, their lives changed. "When you leave that reality of poverty, it is one of the most joyous feelings in the world," she told Bill Moyers, a TV political news commentator and former White House Press Secretary.

It shaped her life—she's dedicated her career to making the world a better place, including environmental, anti-poverty, and justice goals. Her

career includes heading Green For All, a nonprofit that fights for economic justice and a clean-energy economy, and Honor, a home-care startup that helps the elderly get the care they need. She even managed The Artist Formerly Known as Prince. Okay, Prince.

At Green For All, she met Diana, a defense attorney and cofounder of the Ella Baker Center for Human Rights. Their shared values brought them close; as a POC in America, Phaedra knew many African American communities that suffered from mass incarceration, and Diana fought injustice in the courtroom. They founded Promise to improve the criminal justice system.

They started by tackling bail reform; it seemed like such an obvious abuse of power. "The huge number of pretrial individuals in custody, simply because they cannot pay for their release, stood out as a necessary part of our work," Frappier said. Numbers don't lie. Around 2.3 million Americans are in jail or prison, with 4.5 million more on probation or parole. But there are about 8.3 million people a year stuck in local jails who've been charged with no crime, apart from being too poor to pay bail.

Phaedra tapped Jay-Z, whom she knew through Prince, to invest in Promise through his agency, Roc Nation. "Phaedra is building an app that can

help provide 'liberty and justice for all' to millions," he told people—he's been an advocate for ending incarceration for some time now. One in nine Black kids has a parent locked up—and that hurts families. "We can't fix our broken criminal justice system until we take on the exploitative bail industry," he wrote in a *Time* magazine op-ed. "It's time for an innovative and progressive technology that offers sustainable solutions to tough problems." Jay-Z's support helped them raise four million dollars to get Promise off the ground. The app gives users personalized data, including reminders for court dates; case management; and a display of their progress (and compliance), so officers know what's working. Visualizing data like this can also help people on parole with better decision-making.

Justice comes in many forms. In 2016, Laura Montoya, a Latinx engineer from Oakland, founded Accel.AI, an AI development company, to train POC in AI skills. Her focus is on the social and ethical impact of AI tools on the future—which includes building nonbiased systems. That's a really necessary goal—in 2018, tools used by US judges were found to be biased against people of color, and HP's 2009 imaging software couldn't recognize Asian faces. "Technology reflects the people who build it," she said—the more diverse the designers, the better

for everyone. As a second-generation Colombian immigrant, this is how she shows her new country that she cares.

None of these startups have solved the criminal justice problem yet. It's too big a problem to be fixed by one, or even one hundred, companies. But little by little, they are making a difference. They're showing people on the outside the reality of what goes on inside. "It's easy to dehumanize people," Clementine said—the cases that make headlines are normally the worst ones. "Mother of four goes to jail for three unpaid parking tickets" is less common in the newspaper, but is the reality behind bars.

In September 2019, the state of North Dakota officially announced they'd partnered with Recidiviz to track the effectiveness of their policies and processes. "By harnessing this technology, we can enhance public safety, improve lives, and save taxpayer dollars," North Dakota Governor Doug Burgum announced. Their big focus is on keeping people who leave jail from returning—staff now receive daily reports on how their programs are doing. "Increasing reentry success is a win-win-win: It means fewer crimes are being committed, the lives and communities of returning citizens are being improved, and fewer taxpayer dollars are

being spent on incarceration," announced Director Leann Bertsch.

And Recidiviz is growing—in January 2020, they handled data covering 9.5 percent of US incarcerations, and they have more things coming. Women make up 62 percent of their team—"They were the best people we could find!"—and together, they're focused on making a big impact.

"We're a small piece of a big puzzle," Clementine said. "It takes a village to raise a nonprofit. But I feel optimistic about the future of accountability, transparency, and innovation in our criminal justice system."

* * *

We won't get anywhere as a society if we don't start to change the way we think about people who make mistakes. Almost everyone deserves a second chance in life. Don't just think it—say it. Act on it. They'll need help making it happen. That starts here, with you.

In ten years, Recidiviz or Promise or Accel.AI might not be around anymore. But I guarantee that Clementine and Phaedra and Diana and Laura will be, and they'll be working toward the same goals of justice and fairness and equality. And for the people whose lives have been helped by their

companies, they've been everything they needed to be and more.

CHAPTER 6

THE WOMEN WHO INSPIRED US

I would be remiss if this book didn't acknowledge and celebrate the work by women who undoubtedly inspired some of the brilliant minds in this book.

I'm talking about the women who refused to sit on the sidelines, who wouldn't let the patriarchy have the final say, and who pushed forward for science and attacked problems that seemed impossible. In the next two chapters, I'll introduce you to women you might already be aware of, but who deserve to have their successes told again and again and *again.*

Here's to the women who forced their way into our history books—and how the world is better for having had them in it.

Laura Bassi

On October 31, 1711, two hundred years before Marie Curie won the Nobel Prize, Italian scientist Laura Maria Catarina Bassi was born. Bassi was fascinated by science and made it her mission to get the best education possible. She was the second woman in Europe to receive a university degree and the first woman in Europe to be hired by a university. In her role as a physics professor, she introduced Isaac Newton's ideas to Italy. Her work was so impressive that Pope Benedict XIV made a special dispensation

to include her in his twenty-five-person scholar group; she was the only woman allowed. Laura Bassi was an ahead-of-her-time rock star and an inspiration to countless girls in Italy and beyond.

Marie Curie

Marie Curie was born in Warsaw, Poland, on November 7, 1867. She spent her childhood soaking up as much knowledge as possible, and in her teens she attended "the Flying University," a secret school for women who weren't allowed to enroll in university. They met in secret, at a different location each time, to evade the police.

In 1891, she moved to Paris, where she studied physics and the mathematical sciences at the Sorbonne. Here, she met physics professor Pierre Curie, whom she married in 1895. Marie and Pierre worked together to understand radioactivity, and the pair were awarded a joint Nobel Prize in physics for their research in 1903. Marie went on to discover two elements, polonium, which she named after Poland, and radium, after the rays that the element emitted. She was awarded a second Nobel Prize, this time for chemistry, in 1911. Her pioneering work has led to numerous treatments for cancer and has helped shape the world as we know it.

Sophie Germain

Born in Paris, France, on April 1, 1776, Marie-Sophie Germain grew up during the French Revolution. With France in chaos, her pursuits were limited, which gave her time to devour her father's library. Little by little her knowledge expanded, and she found herself drawn to science. She studied Latin and Greek texts so she could read about the work of Isaac Newton. Her parents didn't support her love of higher learning, claiming it was a role best left for men. They tried to discourage her by keeping her room cold at night and limiting her blankets, so she'd lack the energy to learn. But Marie-Sophie was determined, and, while they slept, she'd huddle in some old quilts, light a small candle, and practice her mathematics.

When she turned eighteen, she requested the lecture notes from the École Polytechnique, the French university. She was fascinated by the work of Joseph Louis Lagrange, a renowned astronomer and mathematician. Knowing he wouldn't converse with a girl, she created an alias and started sending him letters. Lagrange was impressed by her insight, and, when he learned she was a girl, he accepted her as a pupil—regardless of her gender, he valued her intelligence. Germain went on to make major contributions to the advancement of mathematics and is best remembered for her work in number theory and Fermat's last theorem.

Maria Margaretha Kirch

Maria Margaretha Kirch, born February 25, 1670, inherited her passion for numbers from her father, a Lutheran minister, who believed that his daughter deserved the same education as any boy from his church. In her early teens, she began studying with Christoph Arnold, a highly respected astronomer who lived nearby. Arnold introduced Maria to German astronomer and mathematician Gottfried Kirch, whom she married. Kirch continued her education, and Maria moved from an apprentice role to that of an assistant (and wife). Together, they made observations and performed calculations to produce calendars and star charts. Every night, starting at nine o'clock, Maria liked to stare at the sky, marveling at the size of the cosmos. During one observation, she noticed the "comet of 1702." She's the first woman in history to find a comet. When her husband died, the university refused to allow her to continue astronomical work, so she kept her work private. Today, she's best known for her discovery of the comet, and for her careful observations of Saturn, Venus, Jupiter, and the sun from 1709 to 1712.

* * *

These are just a few of the women who have proudly carried the science torch and propelled humanity into the future.

CHAPTER 7

THE FUTURE IS NOW

Like the role models in the previous chapter, the women highlighted below are the building blocks of our future. Their brilliant minds are helping shape the way we view medicine, the planet, and the world at large. The earlier chapters introduced some of the most exciting women working in food, robotics, and justice...that you'd probably never heard of. The women featured here are household names; their voices have made the fields of entrepreneurship, leadership, and science more accessible for all.

Hasini Jayatilaka

Hasini Jayatilaka's TED talk on cancer strikes every emotional chord in the human body. The Sri Lankan deftly navigates the roller coaster of fear and anxiety that comes when a family member goes toe-to-toe with this deadly disease. At the end of her talk, she offers some unexpected encouragement. Jayatilaka explains that all humans have a superpower that helps fight cancer, one that makes them stronger than any threat imaginable. "For us humans, collaboration is a superpower that has produced incredible discoveries in the medical and scientific fields," she said.

During her postdoctoral research at Stanford University School of Medicine, Jayatilaka and her team discovered a signal-like pathway that controls how cancer cells spread throughout the body. She developed ways to block these signals, slowing the metastasizing process and improving her patient's survival rates. She's proud of her accomplishments but stresses that everyone on her team deserves credit. "Collaborating is a superpower that we can all turn to," she said. "[It] inspires us to create something bigger than ourselves that will help make the world a better place."

Nasrin Mostafazadeh

Nasrin Mostafazadeh is a scientist who is seeking to help us better understand robotics and artificial intelligence. Growing up, Mostafazadeh's favorite thing to do was to sit in front of her computer and learn how to program. She loved the idea of using software toys to help humanity accomplish great things. "To me, the idea of building a piece of software that automates what you do was really intriguing." Today, she is a senior AI research scientist at Elemental Cognition, where she works on developing AI systems that understand the complexities of human language to such a degree that they can use "common sense" to explain why

and how certain things have happened. Her goal is to use this tool to accelerate scientific discovery for major diseases. "I've chosen to work on the topics of AI that I found to be really challenging in terms of the amount of work that is still needed to be done to even scratch the surface."

Autumn Peltier

Autumn Peltier is a fifteen-year-old teenage activist, fighting for the basic human right to clean water. A member of the Wiikwemkoong First Nation in northern Ontario, she grew up next to one of the world's most amazing freshwater lakes, Lake Huron. But she soon realized that she was one of the lucky ones, as many places in the world around her struggled to provide clean water to their citizens. Autumn decided it was time for her to stand up and help. Taking inspiration from her aunt, Josephine Mandamin, who spent her life raising awareness for water conservation, Peltier began advocating for the universal right to clean drinking water when she was eight years old. At age eleven, Peltier met with Prime Minister Trudeau to speak about pipeline projects the prime minister had approved; these pipelines would impact the water that her tribe, and others like it, would receive. Following this, she traveled to the UN General Assembly in New York, where she

spoke on water rights as part of the International Decade for Action on Water for Sustainable Development. "Water is the lifeblood of Mother Earth...our water should not be for sale. We all have a right to this water as we need it."

Mari Copeny

Mari Copeny, or, as the world has come to know her, "Little Miss Flint," is a thirteen-year-old water activist, fighting to get clean water for the families of Flint, Michigan. Growing up in a world where the only access to fresh water came through water bottles, Mari knew that things had to change. But she was confused by politics and the inability of her so-called leadership to help her family. To get help, Mari made her way to Washington, DC, to attend a congressional hearing on the water crisis. She also wrote a letter to President Obama, asking him to visit Flint in the hopes that he would fully understand the Flint water crisis. She knew that the chances of the president receiving her letter were astronomically low. Six weeks after the hearing, Mari's mom received a call from President Obama, who had received her letter and wanted to visit her in Flint.

President Obama's visit to Flint received worldwide coverage, and Mari collected over $500,000 in donations. That money has impacted the quality of life of some 25,000 children in Flint and beyond. Through these donations and her continued efforts to promote clean water for children around the world, Mari has been able to gather school supplies, toys, bikes, clean water, and other essentials needed to ensure a fulfilled and healthy life for those in need. Most recently, she partnered with a water filtration company to bring state-of-the-art water filters to US communities that are dealing with toxic water.

Greta Thunberg

At first glance, Greta Thunberg looks like an ordinary teenage girl. She is not a scientist. She is not a politician. She doesn't have a college degree. But what she does have is a voice. And her voice has become the voice of a generation and, to some extent, the voice of our planet. She was diagnosed with Asperger syndrome when she was eight years old and her condition sharpened her focus on the problems facing the planet as a whole. "If I were like everyone else, I would have continued on and not seen this crisis," she told *Time* magazine during an interview in 2019, crediting her condition with granting her the ability to "sit for

hours and read things I'm interested in." Frustrated with the misinformation online and the lack of acknowledgment from political leaders, Greta made it her mission to shift the way the world sees the climate crisis. Her passion and candor have helped make her a shining beacon in the climate community, connecting her to world leaders and giving her a platform to speak about the changes that need to happen to build a better future. She's worried that many won't get one, but she's also hopeful that people like her—other smart, brilliant minds, who care about the bigger picture—can change the world for the better.

* * *

As with any list, there are many more amazing women that deserve recognition. Every day, that list gets longer, as young and inquiring minds realize how much they have to give in the science space. Maybe they're building their first science-fair project or singing songs on TikTok to raise awareness about climate change. This book could be ten times as long, and I still couldn't fit even 1 percent of these women in.

FINAL THOUGHTS

"Be less curious about people and more curious about ideas."

—MARIE CURIE

"We realize the importance of our voices only when we are silenced."

—MALALA YOUSAFZAI

It's hard to know what the world will look like in ten years. Technology is moving so fast right now, and it's only going to get faster. It seems every other week our scientists discover new and amazing things. This book provides just a small sample of the many, many women doing awesome world-changing things, in pursuit of a better future for everyone on the planet. #epic

Think of hospitals staffed by friendly robots, people chowing down on delicious "shrimp" and "fish" that started life in a petri dish, and every boy, girl, or gender-nonconforming kid growing up with code-your-own playsets. Our justice system will be fairer, with BIPOC no longer disproportionately charged, and vulnerable people will be provided with mental health resources and work training to help them get back on track. We'll (hopefully) get to zero CO_2 factory emissions, with pollution instantly turned into carbon and biodegradable plastics. People from all economic backgrounds will breathe clean air in their neighborhoods, and robots will do all the nasty, dangerous jobs, freeing up time for people to focus on what's important to them. This isn't wishful thinking—this is scaling up *what we have right now.* (And that's just the people in this book.)

We all have a part to play in shaping the future, regardless of age.

This is your time. Young people have the most power they've ever had, in all of *history*.

Already, tweens and teenagers are changing policy on a global level—activists like Greta Thunberg, Emma González, and Alexandria Villaseñor are marching for what they believe in, telling governments to listen to the scientists, and filing complaints against countries that violate the UN Convention on the Rights of the Child.

Thanks to their impassioned stance on climate change, gun control, and science, they're getting one-on-ones with President Obama, Zendaya, Reese Witherspoon, Ellen DeGeneres, Emma Watson, Oprah, and more. We're not that far off

from a female US president—many other countries have had a female leader by now, and women are setting the agenda worldwide. We've still got a lot to work on—we need more women in senior roles, more diversity in tech and science, and more representation. But I know we can get there—and you are part of that equation. #gucci

In the words of Emma Watson, "Girls should never be afraid to be smart."

Please Stay in Touch!

Find me on:

Instagram @almostzara

Twitter @almostzara

Facebook www.facebook.com/AlmostZara

Website www.zarastone.net

P.S.

I sincerely hope that you enjoyed this book. This work is a labor of love. I wanted to showcase the many different ways girls, gender-nonconforming kids, and boys (we need strong male allies) can be involved in STEAM and the many different pathways into this field: cooking, robotics, nail art...and on and on.

You can find extra details (and pictures and videos) of these kickass women on my website zarastone.net/thefutureofscienceisfemale and on my Facebook page www.facebook.com/thefutureofscienceisfemale.

All feedback is welcome! If you like this book, please let me know (zara@zarastone.net or @almostzara on Twitter and Instagram) and post a review on Amazon. If you don't like it, tell me why—my DMs are always open.

Thanks for reading!

THANKS!

"I do not belong to anyone but myself and neither do you."

—SIMONE GIERTZ, YOUTUBE MAKER

I want to thank all the awesome scientists for their generosity with their time, stories, and labs; you inspired and guided this book, and it wouldn't have been possible without your outstanding work. Your achievements helped make this book what it is.

I couldn't have gotten here without the unbelievable support, friendship, and many, many, tubs of ice cream from my friends, colleagues, and my San Francisco #writerpod. Your insight, feedback, and willingness to be my sounding boards were invaluable in shaping the language, nuances, and beats of these stories.

Thank you: Ellen Airhart, Larissa Zimberoff, Daniela Blei, Nicole Reamey, Juliette Jardim, Kulsum "Cookie" Vakharia-Bodorff, Gabriel Boddorff (for the smiles), Meg Waltner, Mor Goldberger, Nathan Hurst, Sonia Paul, Dr. Harman Boparai, Christina Ku, Toby Eirich, Jodi Eirich, Margie Woodring, Tim Woodring, Ann Young, Bryony Hewer, Becky Dixon,

Robert Markham, Lillian Rafii, Adina Ben-Arie, Gil Ben-Arie, Lavie "the lion cub" Ben-Arie, Jack Rabinowicz, Danielle Rabinowicz, Simon Rabinowicz, Lisa Shaverin, and many, many others. Additional thanks to the San Francisco Writers Grotto for being such a great space to write this. I also want to share my appreciation for the kickass work from my editor Hugo Villabona, for his sensitive feedback and guidance, and for seeing the value in this book, as well as the whole Mango Publishing team for being my cheerleaders every step of this journey. And I'm in awe of artist Jermaine Lau who created the amazing cover art! And finally, I want to thank Thad Eirich and Zero Two for always being awesome; without your support, this work would not have been the same.

Most of all, I want to thank *you*, my readers, for caring, dreaming, and striving.

BIBLIOGRAPHY

The majority of this work comes from interviews I did with scientists and experts. The rest is sourced from scientific papers, media interviews, archives, and books.

Media Outlets

CNN, *Grist* magazine, *Scientific American*, *Fast Company*, National Geographic, *US News*, *The Guardian*, Marketplace, the *Wall Street Journal*, *Essence*, *Vice*, TreeHugger, the *San Francisco Chronicle*, Thomson Reuters Foundation, the *New York Times*, *Forbes*, NPR, Heavy, The Ella Project, Sweety High, *Pacific Standard*, *Quartz News*, NBC DFW, *Wired* magazine, the *Washington Post*, ABC News, *Time* magazine, *Popular Mechanics*, *Popular Science*.

Specialist Media Outlets

Registered Nursing, Nurse.org, *Scrubs* magazine, *PMQ Pizza Magazine*, Deli Market News, Governing, Architect, Global Toy News, PlayMonster, the Strong National Museum of Play, Startland News, Ensia, BusinessGreen, Robotics Business Review, Green

Tech Media, AgFunder News, *Berkeley Science Review*, *Do the Green Thing*, *PCMag*, GreenBiz, Cheddar, The Spoon, *MIT Technology Review*, *IEEE Spectrum*, *The Stanford Daily*, TechCrunch, ScienceDaily, TriplePundit, Republic, Quillette, Uncubed, Scenester.

Public-Sector Organizations

The United Nations Department of Economic and Social Affairs, Food and Agriculture Organization of the United Nations, the National Science Foundation, the World Health Organization, the United States Environmental Protection Agency, NOAA Climate.gov, the National Institute of Corrections.

Independent Research

Oceana, Asthma and Allergy Foundation of America, the Earth Institute of Columbia University, the International Union for Conservation of Nature, the Pantas and Ting Sutardja Center for Entrepreneurship & Technology, Scripps Institution of Oceanography at UC San Diego, TomKat Center for Sustainable Energy, the Center for Data Innovation, the Freshwater Trust, Hotwire Global, Built By Girls, MakerGirl.us, Girls InTech.

Scientific Journals

Nature, the *American Journal of Tropical Medicine and Hygiene*, the Natural Resources Defense Council, the Union of Concerned Scientists, *BMC Health Services Research*, *Stat*, *Pearls*, the Royal Society of Biological Science, *Journal of the American Chemical Society*, *Journal of Thoracic Diseases.*

Books

Aaseng, Nathan. *Business Builders in Fast Food*. Oliver Press, 2001.

Heather Cabot, Walravens, Samantha. *Geek Girl Rising: Inside the Sisterhood Shaking Up Tech.* St. Martin's Publishing Group, 2017.

Bokova, Irina. *Cracking the Code: Girls' and Women's Education in Science, Technology, Engineering and Mathematics.* The United Nations Educational, Scientific and Cultural Organization. 2017.

F. Bailey Norwood, Tamara L. Mix. *Meet the Food Radicals.* Oxford University Press. 2019.

ABOUT THE AUTHOR

Zara Stone is an award-winning journalist who has been published and has appeared in *The Atlantic*, the *Washington Post*, *Vice*, *Forbes*, *Wired*, *The Wall Street Journal*, the *New York Post*, *ABC News*, the *BBC*, *COZY* Media, *Atlas Obscura*, *Cosmopolitan* magazine, *NPR*, *OneZero*, *Elemental*, *Marker*, and *BuzzFeed*, amongst others.

As a reporter she covers technology and culture and the eccentricities of the startup world. *The Future of Science Is Female* is her first book. She was born in London, England, and moved to the US for graduate school. She lives in San Francisco.

mango
PUBLISHING GROUP
TM